FREE Test Taking Tips DVD Offer

To help us better serve you, we have developed a Test Taking Tips DVD that we would like to give you for FREE. **This DVD covers world-class test taking tips that you can use to be even more successful when you are taking your test.**

All that we ask is that you email us your feedback about your study guide. Please let us know what you thought about it – whether that is good, bad or indifferent.

To get your **FREE Test Taking Tips DVD**, email freedvd@studyguideteam.com with "FREE DVD" in the subject line and the following information in the body of the email:

    a. The title of your study guide.

    b. Your product rating on a scale of 1-5, with 5 being the highest rating.

    c. Your feedback about the study guide. What did you think of it?

    d. Your full name and shipping address to send your free DVD.

If you have any questions or concerns, please don't hesitate to contact us at freedvd@studyguideteam.com.

Thanks again!

TACHS Exam Study Guide

TACHS Test Prep and Practice Test Questions for the Catholic High School Entrance Exam [2nd Edition]

TPB Publishing

Interested in buying more than 10 copies of our product? Contact us about bulk discounts:
bulkorders@studyguideteam.com

ISBN 13: 9781628456646
ISBN 10: 1628456647

Table of Contents

Quick Overview

As you draw closer to taking your exam, effective preparation becomes more and more important. Thankfully, you have this study guide to help you get ready. Use this guide to help keep your studying on track and refer to it often.

This study guide contains several key sections that will help you be successful on your exam. The guide contains tips for what you should do the night before and the day of the test. Also included are test-taking tips. Knowing the right information is not always enough. Many well-prepared test takers struggle with exams. These tips will help equip you to accurately read, assess, and answer test questions.

A large part of the guide is devoted to showing you what content to expect on the exam and to helping you better understand that content. In this guide are practice test questions so that you can see how well you have grasped the content. Then, answer explanations are provided so that you can understand why you missed certain questions.

Don't try to cram the night before you take your exam. This is not a wise strategy for a few reasons. First, your retention of the information will be low. Your time would be better used by reviewing information you already know rather than trying to learn a lot of new information. Second, you will likely become stressed as you try to gain a large amount of knowledge in a short amount of time. Third, you will be depriving yourself of sleep. So be sure to go to bed at a reasonable time the night before. Being well-rested helps you focus and remain calm.

Be sure to eat a substantial breakfast the morning of the exam. If you are taking the exam in the afternoon, be sure to have a good lunch as well. Being hungry is distracting and can make it difficult to focus. You have hopefully spent lots of time preparing for the exam. Don't let an empty stomach get in the way of success!

When travelling to the testing center, leave earlier than needed. That way, you have a buffer in case you experience any delays. This will help you remain calm and will keep you from missing your appointment time at the testing center.

Be sure to pace yourself during the exam. Don't try to rush through the exam. There is no need to risk performing poorly on the exam just so you can leave the testing center early. Allow yourself to use all of the allotted time if needed.

Remain positive while taking the exam even if you feel like you are performing poorly. Thinking about the content you should have mastered will not help you perform better on the exam.

Once the exam is complete, take some time to relax. Even if you feel that you need to take the exam again, you will be well served by some down time before you begin studying again. It's often easier to convince yourself to study if you know that it will come with a reward!

Test-Taking Strategies

1. Predicting the Answer

When you feel confident in your preparation for a multiple-choice test, try predicting the answer before reading the answer choices. This is especially useful on questions that test objective factual knowledge. By predicting the answer before reading the available choices, you eliminate the possibility that you will be distracted or led astray by an incorrect answer choice. You will feel more confident in your selection if you read the question, predict the answer, and then find your prediction among the answer choices. After using this strategy, be sure to still read all of the answer choices carefully and completely. If you feel unprepared, you should not attempt to predict the answers. This would be a waste of time and an opportunity for your mind to wander in the wrong direction.

2. Reading the Whole Question

Too often, test takers scan a multiple-choice question, recognize a few familiar words, and immediately jump to the answer choices. Test authors are aware of this common impatience, and they will sometimes prey upon it. For instance, a test author might subtly turn the question into a negative, or he or she might redirect the focus of the question right at the end. The only way to avoid falling into these traps is to read the entirety of the question carefully before reading the answer choices.

3. Looking for Wrong Answers

Long and complicated multiple-choice questions can be intimidating. One way to simplify a difficult multiple-choice question is to eliminate all of the answer choices that are clearly wrong. In most sets of answers, there will be at least one selection that can be dismissed right away. If the test is administered on paper, the test taker could draw a line through it to indicate that it may be ignored; otherwise, the test taker will have to perform this operation mentally or on scratch paper. In either case, once the obviously incorrect answers have been eliminated, the remaining choices may be considered. Sometimes identifying the clearly wrong answers will give the test taker some information about the correct answer. For instance, if one of the remaining answer choices is a direct opposite of one of the eliminated answer choices, it may well be the correct answer. The opposite of obviously wrong is obviously right! Of course, this is not always the case. Some answers are obviously incorrect simply because they are irrelevant to the question being asked. Still, identifying and eliminating some incorrect answer choices is a good way to simplify a multiple-choice question.

4. Don't Overanalyze

Anxious test takers often overanalyze questions. When you are nervous, your brain will often run wild, causing you to make associations and discover clues that don't actually exist. If you feel that this may be a problem for you, do whatever you can to slow down during the test. Try taking a deep breath or counting to ten. As you read and consider the question, restrict yourself to the particular words used by the author. Avoid thought tangents about what the author *really* meant, or what he or she was *trying* to say. The only things that matter on a multiple-choice test are the words that are actually in the question. You must avoid reading too much into a multiple-choice question, or supposing that the writer meant something other than what he or she wrote.

5. No Need for Panic

It is wise to learn as many strategies as possible before taking a multiple-choice test, but it is likely that you will come across a few questions for which you simply don't know the answer. In this situation, avoid panicking. Because most multiple-choice tests include dozens of questions, the relative value of a single wrong answer is small. As much as possible, you should compartmentalize each question on a multiple-choice test. In other words, you should not allow your feelings about one question to affect your success on the others. When you find a question that you either don't understand or don't know how to answer, just take a deep breath and do your best. Read the entire question slowly and carefully. Try rephrasing the question a couple of different ways. Then, read all of the answer choices carefully. After eliminating obviously wrong answers, make a selection and move on to the next question.

6. Confusing Answer Choices

When working on a difficult multiple-choice question, there may be a tendency to focus on the answer choices that are the easiest to understand. Many people, whether consciously or not, gravitate to the answer choices that require the least concentration, knowledge, and memory. This is a mistake. When you come across an answer choice that is confusing, you should give it extra attention. A question might be confusing because you do not know the subject matter to which it refers. If this is the case, don't eliminate the answer before you have affirmatively settled on another. When you come across an answer choice of this type, set it aside as you look at the remaining choices. If you can confidently assert that one of the other choices is correct, you can leave the confusing answer aside. Otherwise, you will need to take a moment to try to better understand the confusing answer choice. Rephrasing is one way to tease out the sense of a confusing answer choice.

7. Your First Instinct

Many people struggle with multiple-choice tests because they overthink the questions. If you have studied sufficiently for the test, you should be prepared to trust your first instinct once you have carefully and completely read the question and all of the answer choices. There is a great deal of research suggesting that the mind can come to the correct conclusion very quickly once it has obtained all of the relevant information. At times, it may seem to you as if your intuition is working faster even than your reasoning mind. This may in fact be true. The knowledge you obtain while studying may be retrieved from your subconscious before you have a chance to work out the associations that support it. Verify your instinct by working out the reasons that it should be trusted.

8. Key Words

Many test takers struggle with multiple-choice questions because they have poor reading comprehension skills. Quickly reading and understanding a multiple-choice question requires a mixture of skill and experience. To help with this, try jotting down a few key words and phrases on a piece of scrap paper. Doing this concentrates the process of reading and forces the mind to weigh the relative importance of the question's parts. In selecting words and phrases to write down, the test taker thinks about the question more deeply and carefully. This is especially true for multiple-choice questions that are preceded by a long prompt.

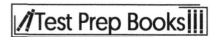

9. Subtle Negatives

One of the oldest tricks in the multiple-choice test writer's book is to subtly reverse the meaning of a question with a word like *not* or *except*. If you are not paying attention to each word in the question, you can easily be led astray by this trick. For instance, a common question format is, "Which of the following is...?" Obviously, if the question instead is, "Which of the following is not...?," then the answer will be quite different. Even worse, the test makers are aware of the potential for this mistake and will include one answer choice that would be correct if the question were not negated or reversed. A test taker who misses the reversal will find what he or she believes to be a correct answer and will be so confident that he or she will fail to reread the question and discover the original error. The only way to avoid this is to practice a wide variety of multiple-choice questions and to pay close attention to each and every word.

10. Reading Every Answer Choice

It may seem obvious, but you should always read every one of the answer choices! Too many test takers fall into the habit of scanning the question and assuming that they understand the question because they recognize a few key words. From there, they pick the first answer choice that answers the question they believe they have read. Test takers who read all of the answer choices might discover that one of the latter answer choices is actually *more* correct. Moreover, reading all of the answer choices can remind you of facts related to the question that can help you arrive at the correct answer. Sometimes, a misstatement or incorrect detail in one of the latter answer choices will trigger your memory of the subject and will enable you to find the right answer. Failing to read all of the answer choices is like not reading all of the items on a restaurant menu: you might miss out on the perfect choice.

11. Spot the Hedges

One of the keys to success on multiple-choice tests is paying close attention to every word. This is never truer than with words like almost, most, some, and sometimes. These words are called "hedges" because they indicate that a statement is not totally true or not true in every place and time. An absolute statement will contain no hedges, but in many subjects, the answers are not always straightforward or absolute. There are always exceptions to the rules in these subjects. For this reason, you should favor those multiple-choice questions that contain hedging language. The presence of qualifying words indicates that the author is taking special care with his or her words, which is certainly important when composing the right answer. After all, there are many ways to be wrong, but there is only one way to be right! For this reason, it is wise to avoid answers that are absolute when taking a multiple-choice test. An absolute answer is one that says things are either all one way or all another. They often include words like *every, always, best,* and *never.* If you are taking a multiple-choice test in a subject that doesn't lend itself to absolute answers, be on your guard if you see any of these words.

12. Long Answers

In many subject areas, the answers are not simple. As already mentioned, the right answer often requires hedges. Another common feature of the answers to a complex or subjective question are qualifying clauses, which are groups of words that subtly modify the meaning of the sentence. If the question or answer choice describes a rule to which there are exceptions or the subject matter is complicated, ambiguous, or confusing, the correct answer will require many words in order to be expressed clearly and accurately. In essence, you should not be deterred by answer choices that seem excessively long. Oftentimes, the author of the text will not be able to write the correct answer without

offering some qualifications and modifications. Your job is to read the answer choices thoroughly and completely and to select the one that most accurately and precisely answers the question.

13. Restating to Understand

Sometimes, a question on a multiple-choice test is difficult not because of what it asks but because of how it is written. If this is the case, restate the question or answer choice in different words. This process serves a couple of important purposes. First, it forces you to concentrate on the core of the question. In order to rephrase the question accurately, you have to understand it well. Rephrasing the question will concentrate your mind on the key words and ideas. Second, it will present the information to your mind in a fresh way. This process may trigger your memory and render some useful scrap of information picked up while studying.

14. True Statements

Sometimes an answer choice will be true in itself, but it does not answer the question. This is one of the main reasons why it is essential to read the question carefully and completely before proceeding to the answer choices. Too often, test takers skip ahead to the answer choices and look for true statements. Having found one of these, they are content to select it without reference to the question above. Obviously, this provides an easy way for test makers to play tricks. The savvy test taker will always read the entire question before turning to the answer choices. Then, having settled on a correct answer choice, he or she will refer to the original question and ensure that the selected answer is relevant. The mistake of choosing a correct-but-irrelevant answer choice is especially common on questions related to specific pieces of objective knowledge. A prepared test taker will have a wealth of factual knowledge at his or her disposal, and should not be careless in its application.

15. No Patterns

One of the more dangerous ideas that circulates about multiple-choice tests is that the correct answers tend to fall into patterns. These erroneous ideas range from a belief that B and C are the most common right answers, to the idea that an unprepared test-taker should answer "A-B-A-C-A-D-A-B-A." It cannot be emphasized enough that pattern-seeking of this type is exactly the WRONG way to approach a multiple-choice test. To begin with, it is highly unlikely that the test maker will plot the correct answers according to some predetermined pattern. The questions are scrambled and delivered in a random order. Furthermore, even if the test maker was following a pattern in the assignation of correct answers, there is no reason why the test taker would know which pattern he or she was using. Any attempt to discern a pattern in the answer choices is a waste of time and a distraction from the real work of taking the test. A test taker would be much better served by extra preparation before the test than by reliance on a pattern in the answers.

FREE DVD OFFER

Don't forget that doing well on your exam includes both understanding the test content and understanding how to use what you know to do well on the test. We offer a completely FREE Test Taking Tips DVD that covers world class test taking tips that you can use to be even more successful when you are taking your test.

All that we ask is that you email us your feedback about your study guide. To get your **FREE Test Taking Tips DVD**, email freedvd@studyguideteam.com with "FREE DVD" in the subject line and the following information in the body of the email:

- The title of your study guide.
- Your product rating on a scale of 1-5, with 5 being the highest rating.
- Your feedback about the study guide. What did you think of it?
- Your full name and shipping address to send your free DVD.

Introduction to the TACHS Exam

Function of the Test

The Test for Admission into Catholic High Schools (TACHS) is an entrance exam to determine admittance into Catholic high schools in the New York City area, or for the allocation of scholarship funds. These high schools include the Diocese of Brooklyn or Queens, the Archdiocese of New York, or the Diocese of Rockville Centre, along with some independent schools. Students who take this exam are in the eighth grade waiting to be accepted into their freshmen class. The exam was first given in the fall of 2004. In 2017, the number of eighth graders taking the TACHS exam was approximately 1,500.

Test Administration

The test is given each year in November, and the results are expected back in either January or February. Students are only allowed to apply to three high schools per exam. For the testing location, check with your middle school administrators. Only eighth graders may take the exam, so keep in mind that a student may only take the test once. For accommodations, the Archdiocese of New York and the Diocese of Brooklyn/Queens offer an extended length in test time depending on the situation.

Test Format

Students will need their student ID, the Admit Card, and two sharpened No. 2 pencils with erasers on test day. Cell phones, watches, calculators, food, and drink are not allowed.

The TACHS exam consists of four sections with a total of 200 multiple-choice questions. The four sections are Reading, Written Expression, Mathematics, and Abilities. The following table depicts the content as well as the amount of questions each section will have:

Subject	Parts	Questions	Time
Reading	1. Vocabulary 2. Reading Comprehension	20 questions 30 questions	10 minutes 25 minutes
Written Expression	1. Spelling, Capitalization, Punctuation, and Usage/Expression 2. Paragraphs	40 questions 10 questions	23 minutes 7 minutes
Mathematics	1. Concepts, Data Interpretation, and Problem Solving 2. Estimation	32 questions 18 questions	33 minutes 7 minutes
Abilities	1. Figure Matrices 2. Paper Folding 3. Figure Classifications	20 questions 20 questions 23 questions	32 minutes

Scoring

For Catholic schools, once exams are complete, the score reports will arrive at the high schools on December 14 for the 2018 exam reports. Students should receive their score reports at their Catholic Elementary School no later than January 25. For public and private school students, the score reports will be sent out January 16, 2018 to the address provided and should arrive no later than January 23.

Study Prep Plan for the TACHS Exam

1 **Schedule** - Use one of our study schedules below or come up with one of your own.

2 **Relax** - Test anxiety can hurt even the best students. There are many ways to reduce stress. Find the one that works best for you.

3 **Execute** - Once you have a good plan in place, be sure to stick to it.

One Week Study Schedule

Day 1	Reading
Day 2	Written Expression
Day 3	Math
Day 4	Ability
Day 5	Practice Questions
Day 6	Review Answer Explanations
Day 7	Take Your Exam!

Two Week Study Schedule

Day 1	Indentifying Modes of Writing	Day 8	Number Sense and Operations
Day 2	Figurative Language	Day 9	Geometry
Day 3	Text Structure	Day 10	Practice Questions
Day 4	Practice Questions	Day 11	Ability
Day 5	Conventions of Standard English	Day 12	Practice Questions
Day 6	Affixes, Context, and Syntax	Day 13	Practice Questions
Day 7	Practice Questions	Day 14	Take Your Exam!

One Month Study Schedule

Day 1	Supporting Details	Day 11	Identifying the Position and Purpose	Day 21	Data Analysis
Day 2	Identifying Modes of Writing	Day 12	Parts of Speech	Day 22	Probability and Statistics
Day 3	Using Context Clues	Day 13	Errors in Standard English Grammar, Usage, Syntax, and Mechanics	Day 23	Geometry
Day 4	Figurative Language	Day 14	Grammar, Usage, Syntax, and Mechanics Choices	Day 24	Measurement
Day 5	Development of Themes	Day 15	Components of Sentences	Day 25	Estimation
Day 6	Point of View	Day 16	Structure of Sentences	Day 26	Ability
Day 7	Style, Tone, and Mood	Day 17	Affixes, Context, and Syntax	Day 27	Practice Questions
Day 8	Drawing Conclusions	Day 18	Number Sense and Operations	Day 28	Practice Questions
Day 9	Opinions, Facts, and Fallacies	Day 19	Word Problems	Day 29	Review Answer Explanations
Day 10	Effects of Word Choice	Day 20	Algebraic Patterns and Connections	Day 30	Take Your Exam!

Reading

Topic Versus the Main Idea

It is very important to know the difference between the topic and the main idea of the text. Even though these two are similar because they both present the central point of a text, they have distinctive differences. A *topic* is the subject of the text; it can usually be described in a one- to two-word phrase and appears in the simplest form. On the other hand, the *main idea* is more detailed and provides the author's central point of the text. It can be expressed through a complete sentence and is often found in the beginning, middle, or end of a paragraph. In most nonfiction books, the first sentence of the passage usually (but not always) states the main idea. Review the passage below to explore the topic versus the main idea.

Cheetahs

Cheetahs are one of the fastest mammals on the land, reaching up to 70 miles an hour over short distances. Even though cheetahs can run as fast as 70 miles an hour, they usually only have to run half that speed to catch up with their choice of prey. Cheetahs cannot maintain a fast pace over long periods of time because their bodies will overheat. After a chase, cheetahs need to rest for approximately 30 minutes prior to eating or returning to any other activity.

In the example above, the topic of the passage is "Cheetahs" simply because that is the subject of the text. The main idea of the text is "Cheetahs are one of the fastest mammals on the land but can only maintain a fast pace for shorter distances." While it covers the topic, it is more detailed and refers to the text in its entirety. The text continues to provide additional details called *supporting details,* which will be discussed in the next section.

Supporting Details

Supporting details help readers better develop and understand the main idea. Supporting details answer questions like *who, what, where, when, why,* and *how.* Different types of supporting details include examples, facts and statistics, anecdotes, and sensory details.

Persuasive and informative texts often use supporting details. In persuasive texts, authors attempt to make readers agree with their points of view, and supporting details are often used as "selling points." If authors make a statement, they need to support the statement with evidence in order to adequately persuade readers. Informative texts use supporting details such as examples and facts to inform readers. Review the previous "Cheetahs" passage to find examples of supporting details.

Cheetahs

Cheetahs are one of the fastest mammals on the land, reaching up to 70 miles an hour over short distances. Even though cheetahs can run as fast as 70 miles an hour, they usually only have to run half that speed to catch up with their choice of prey. Cheetahs cannot maintain a fast pace over long periods of time because their bodies will overheat. After a chase, cheetahs need to rest for approximately 30 minutes prior to eating or returning to any other activity.

In the example, supporting details include:

1. Cheetahs reach up to 70 miles per hour over short distances.
2. They usually only have to run half that speed to catch up with their prey.
3. Cheetahs will overheat if they exert a high speed over longer distances.
4. Cheetahs need to rest for 30 minutes after a chase.

Look at the diagram below (applying the cheetah example) to help determine the hierarchy of topic, main idea, and supporting details.

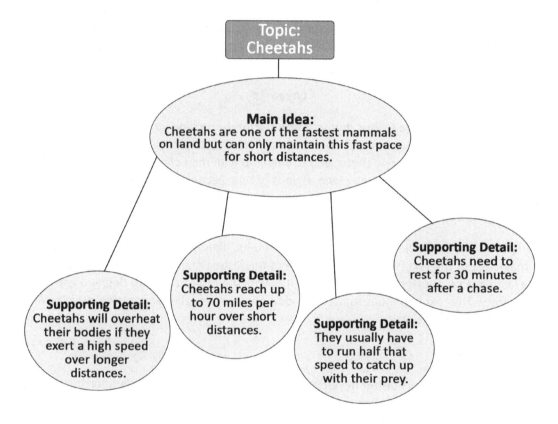

Author's Intent

No matter the genre or format, all authors are writing to persuade, inform, entertain, or express feelings. Often, these purposes are blended, with one dominating the rest. It's useful to learn to recognize the author's intent.

Persuasive writing is used to persuade or convince readers of something. It often contains two elements: the argument and the counterargument. The argument takes a stance on an issue, while the counterargument pokes holes in the opposition's stance. Authors rely on logic, emotion, and writer credibility to persuade readers to agree with them. If readers are opposed to the stance before reading, they are unlikely to adopt that stance. However, those who are undecided or committed to the same stance are more likely to agree with the author.

Informative writing tries to teach or inform. Workplace manuals, instructor lessons, statistical reports and cookbooks are examples of informative texts. Informative writing is usually based on facts and is often void of emotion and persuasion. Informative texts generally contain statistics, charts, and graphs.

Though most informative texts lack a persuasive agenda, readers must examine the text carefully to determine whether one exists within a given passage.

Stories or narratives are designed to entertain. When you go to the movies, you often want to escape for a few hours, not necessarily to think critically. Entertaining writing is designed to delight and engage the reader. However, sometimes this type of writing can be woven into more serious materials, such as persuasive or informative writing to hook the reader before transitioning into a more scholarly discussion.

Emotional writing works to evoke the reader's feelings, such as anger, euphoria, or sadness. The connection between reader and author is an attempt to cause the reader to share the author's intended emotion or tone. Sometimes in order to make a piece more poignant, the author simply wants readers to feel the same emotions that the author has felt. Other times, the author attempts to persuade or manipulate the reader into adopting his stance. While it's okay to sympathize with the author, be aware of the individual's underlying intent.

Identifying Modes of Writing

Distinguishing Between Common Modes of Writing

To distinguish between the common modes of writing, it is important to identify the primary purpose of the work. This can be determined by considering what the author is trying to say to the reader. Although there are countless different styles of writing, all written works tend to fall under four primary categories: argumentative/persuasive, informative expository, descriptive, and narrative.

The table below highlights the purpose, distinct characteristics, and examples of each rhetorical mode.

Writing Mode	Purpose	Distinct Characteristics	Examples
Argumentative	To persuade	Opinions, loaded or subjective language, evidence, suggestions of what the reader should do, calls to action	Critical reviews Political journals Letters of recommendation Cover letters Advertising
Informative	To teach or inform	Objective language, definitions, instructions, factual information	Business and scientific reports Textbooks Instruction manuals News articles Personal letters Wills Informative essays Travel guides Study guides

Writing Mode	Purpose	Distinct Characteristics	Examples
Descriptive	To deliver sensory details to the reader	Heavy use of adjectives and imagery, language that appeals to any of the five senses	Poetry Journal entries Often used in narrative mode
Narrative	To tell a story, share an experience, entertain	Series of events, plot, characters, dialogue, conflict	Novels Short stories Novellas Anecdotes Biographies Epic poems Autobiographies

Identifying Common Types of Writing

The following steps help to identify examples of common types within the modes of writing:

1. Identifying the audience—to whom or for whom the author is writing
2. Determining the author's purpose—why the author is writing the piece
3. Analyzing the word choices and how they are used

To demonstrate, the following passage has been marked to illustrate *the addressee*, the author's purpose, and word choices:

> *To Whom It May Concern*:
>
> I am extraordinarily excited to be applying to the Master of Environmental Science program at Australian National University. I believe the richness in biological and cultural diversity, as well as Australia's close proximity to the Great Barrier Reef, would provide a deeply fulfilling educational experience. *I am writing to express why I believe I would be an excellent addition to the program.*
>
> While in college, I participated in a three-month public health internship in Ecuador, where I spent time both learning about medicine in a third world country and also about the Ecuadorian environment, including the Amazon Jungle and the Galápagos Islands. My favorite experience through the internship, besides swimming with sea lions in San Cristóbal, was helping to neutralize parasitic potable water and collect samples for analysis in Puyo.
>
> Though my undergraduate studies related primarily to the human body, I took several courses in natural science, including a year of chemistry, biology, and physics as well as a course in a calculus. I am confident that my fundamental knowledge in these fields will prepare me for the science courses integral to the Masters of Environmental Science.

Having identified the *addressee*, it is evident that this selection is a letter of some kind. Further inspection into the author's purpose, seen in *bold*, shows that the author is trying to explain why he or she should be accepted into the environmental science program, which automatically places it into the argumentative mode as the writer is trying to persuade the reader to agree and to incite the reader into action by encouraging the program to accept the writer as a candidate. In addition to revealing the purpose, the use of emotional language—extraordinarily, excellent, deeply fulfilling, favorite experience,

confident—illustrates that this is a persuasive piece. It also provides evidence for why this person would be an excellent addition to the program—his/her experience in Ecuador and with scientific curriculum.

The following passage presents an opportunity to solidify this method of analysis and practice the steps above to determine the mode of writing:

> The biological effects of laughter have long been an interest of medicine and psychology. Laughing is often speculated to reduce blood pressure because it induces feelings of relaxation and elation. Participating students watched a series of videos that elicited laughter, and their blood pressure was taken before and after the viewings. An average decrease in blood pressure was observed, though resulting p-values attest that the results were not significant.

This selection contains factual and scientific information, is devoid of any adjectives or flowery descriptions, and is not trying to convince the reader of any particular stance. Though the audience is not directly addressed, the purpose of the passage is to present the results of an experiment to those who would be interested in the biological effects of laughter—most likely a scientific community. Thus, this passage is an example of informative writing.

Below is another passage to help identify examples of the common writing modes, taken from *The Endeavor Journal of Sir Joseph Banks*:

10th May 1769 – THE ENGLISH CREW GET TAHITIAN NAMES

> We have now got the Indian name of the Island, Otahite, so therefore for the future I shall call it. As for our own names the Indians find so much dificulty in pronouncing them that we are forcd to indulge them in calling us what they please, or rather what they say when they attempt to pronounce them. I give here the List: Captn Cooke *Toote*, Dr Solander *Torano*, Mr Hicks *Hete*, Mr Gore *Toárro*, Mr Molineux *Boba* from his Christian name Robert, Mr Monkhouse *Mato*, and myself *Tapáne*. In this manner they have names for almost every man in the ship.

This extract contains no elements of an informative or persuasive intent and does not seem to follow any particular line of narrative. The passage gives a list of the different names that the Indians have given the crew members, as well as the name of an island. Although there is no context for the selection, through the descriptions, it is clear that the author and his comrades are on an island trying to communicate with the native inhabitants. Hence, this passage is a journal that reflects the descriptive mode.

These are only a few of the many examples that can be found in the four primary modes of writing.

Determining the Appropriate Mode of Writing

The author's *primary purpose* is defined as the reason an author chooses to write a selection, and it is often dependent on his or her *audience*. A biologist writing a textbook, for example, does so to communicate scientific knowledge to an audience of people who want to study biology. An audience can be as broad as the entire global population or as specific as women fighting for equal rights in the bicycle repair industry. Whatever the audience, it is important that the author considers its demographics—age, gender, culture, language, education level, etc.

If the author's purpose is to persuade or inform, he or she will consider how much the intended audience knows about the subject. For example, if an author is writing on the importance of recycling to anyone who will listen, he or she will use the informative mode—including background information on recycling—and the argumentative mode—evidence for why it works, while also using simple diction so that it is easy for everyone to understand. If, on the other hand, the writer is proposing new methods for recycling using solar energy, the audience is probably already familiar with standard recycling processes and will require less background information, as well as more technical language inherent to the scientific community.

If the author's purpose is to entertain through a story or a poem, he or she will need to consider whom he/she is trying to entertain. If an author is writing a script for a children's cartoon, the plot, language, conflict, characters, and humor would align with the interests of the age demographic of that audience. On the other hand, if an author is trying to entertain adults, he or she may write content not suitable for children. The author's purpose and audience are generally interdependent.

Using Context Clues

A context clue is a hint that an author provides to the reader in order to help define difficult or unique words. When reading a passage, a test taker should take note of any unfamiliar words, and then examine the sentence around them to look for clues to the word meanings. Let's look at an example:

> He faced a *conundrum* in making this decision. He felt as if he had come to a crossroads. This was truly a puzzle, and what he did next would determine the course of his future.

The word *conundrum* may be unfamiliar to the reader. By looking at context clues, the reader should be able to determine its meaning. In this passage, context clues include the idea of making a decision and of being unsure. Furthermore, the author restates the definition of conundrum in using the word *puzzle* as a synonym. Therefore, the reader should be able to determine that the definition of the word *conundrum* is a difficult puzzle.

Similarly, a reader can determine difficult vocabulary by identifying antonyms. Let's look at an example:

> Her *gregarious* nature was completely opposite of her twin's, who was shy, retiring, and socially nervous.

The word *gregarious* may be unfamiliar. However, by looking at the surrounding context clues, the reader can determine that *gregarious* does not mean shy. The twins' personalities are being contrasted. Therefore, *gregarious* must mean sociable, or something similar to it.

At times, an author will provide contextual clues through a cause and effect relationship. Look at the next sentence as an example:

> The athletes were excited with *elation* when they won the tournament; unfortunately, their off-court antics caused them to forfeit the win.

The word *elation* may be unfamiliar to the reader. However, the author defines the word by presenting a cause and effect relationship. The athletes were so elated at the win that their behavior went overboard and they had to forfeit. In this instance, *elated* must mean something akin to overjoyed, happy, and overexcited.

Cause and effect is one technique authors use to demonstrate relationships. A cause is why something happens. The effect is what happens as a result. For example, a reader may encounter text such as *Because he was unable to sleep, he was often restless and irritable during the day*. The cause is insomnia due to lack of sleep. The effect is being restless and irritable. When reading for a cause and effect

relationship, look for words such as "if", "then", "such", and "because." By using cause and effect, an author can describe direct relationships, and convey an overall theme, particularly when taking a stance on their topic.

An author can also provide contextual clues through comparison and contrast. Let's look at an example:

> Her torpid state caused her parents, and her physician, to worry about her seemingly sluggish well-being.

The word *torpid* is probably unfamiliar to the reader. However, the author has compared *torpid* to a state of being and, moreover, one that's worrisome. Therefore, the reader should be able to determine that *torpid* is not a positive, healthy state of being. In fact, through the use of comparison, it means sluggish. Similarly, an author may contrast an unfamiliar word with an idea. In the sentence *Her torpid state was completely opposite of her usual, bubbly self,* the meaning of *torpid*, or sluggish, is contrasted with the words *bubbly self*.

A test taker should be able to critically assess and determine unfamiliar word meanings through the use of an author's context clues in order to fully comprehend difficult text passages.

Relating Unfamiliar Words to Familiar Words

The Reading section will test a reader's ability to use context clues, and then relate unfamiliar words to more familiar ones. Using the word *torpid* as an example, the test may ask the test taker to relate the meaning of the word to a list of vocabulary options and choose the more familiar word as closest in meaning. In this case, the test may say something like the following:

> Which of the following words means the same as the word *torpid* in the above passage?

Then they will provide the test taker with a list of familiar options such as happy, disgruntled, sluggish, and animated. By using context clues, the reader has already determined the meaning of *torpid* as slow or sluggish, so the reader should be able to correctly identify the word *sluggish* as the correct answer.

One effective way to relate unfamiliar word meanings to more familiar ones is to substitute the provided word answer options with the unfamiliar word in question. Although this will not always lead to a correct answer every time, this strategy will help the test taker narrow answer options. Be careful when utilizing this strategy. Pay close attention to the meaning of sentences and answer choices because it's easy to mistake answer choices as correct when they are easily substituted, especially when they are the same part of speech. Does the sentence mean the same thing with the substituted word option in place or does it change entirely? Does the substituted word make sense? Does it possibly mean the same as the unfamiliar word in question?

Figurative Language

Literary texts also employ rhetorical devices. Figurative language like simile and metaphor is a type of rhetorical device commonly found in literature. In addition to rhetorical devices that play on the *meanings* of words, there are also rhetorical devices that use the *sounds* of words. These devices are most often found in poetry but may also be found in other types of literature and in non-fiction writing like speech texts.

Alliteration and *assonance* are both varieties of sound repetition. Other types of sound repetition include: anaphora, repetition that occurs at the beginning of the sentences; epiphora, repetition occurring at the end of phrases; antimetabole, repetition of words in reverse order; and antiphrasis, a form of denial of an assertion in a text.

Alliteration refers to the repetition of the first sound of each word. Recall Robert Burns' opening line:

> My love is like a red, red rose

This line includes two instances of alliteration: "love" and "like" (repeated *L* sound), as well as "red" and "rose" (repeated *R* sound). Next, assonance refers to the repetition of vowel sounds, and can occur anywhere within a word (not just the opening sound). Here is the opening of a poem by John Keats:

> When I have fears that I may cease to be

> Before my pen has glean'd my teeming brain

Assonance can be found in the words "fears," "cease," "be," "glean'd," and "teeming," all of which stress the long *E* sound. Both alliteration and assonance create a harmony that unifies the writer's language.

Another sound device is *onomatopoeia*, or words whose spelling mimics the sound they describe. Words such as "crash," "bang," and "sizzle" are all examples of onomatopoeia. Use of onomatopoetic language adds auditory imagery to the text.

Readers are probably most familiar with the technique of *pun*. A pun is a play on words, taking advantage of two words that have the same or similar pronunciation. Puns can be found throughout Shakespeare's plays, for instance:

> Now is the winter of our discontent
> Made glorious summer by this son of York

These lines from *Richard III* contain a play on words. Richard III refers to his brother, the newly crowned King Edward IV, as the "son of York," referencing their family heritage from the house of York. However, while drawing a comparison between the political climate and the weather (times of political trouble were the "winter," but now the new king brings "glorious summer"), Richard's use of the word "son" also implies another word with the same pronunciation, "sun"—so Edward IV is also like the sun, bringing light, warmth, and hope to England. Puns are a clever way for writers to suggest two meanings at once.

Some examples of figurative language are included in the following graphic.

Term	Definition	Example
Simile	Compares two things using "like" or "as"	Her hair was like gold.
Metaphor	Compares two things as if they are the same	He was a giant teddy bear.
Idiom	Using words with predictable meanings to create a phrase with a different meaning	The world is your oyster.
Alliteration	Repeating the same beginning sound or letter in a phrase for emphasis	The busy baby babbled.
Personification	Attributing human characteristics to an object or an animal	The house glowered menacingly with a dark smile.
Foreshadowing	Giving an indication that something is going to happen later in the story	I wasn't aware at the time, but I would come to regret those words.
Symbolism	Using symbols to represent ideas and provide a different meaning	The ring represented the bond between us.
Onomatopoeia	Using words that imitate sound	The tire went off with a bang and a crunch.
Imagery	Appealing to the senses by using descriptive language	The sky was painted with red and pink and streaked with orange.
Hyperbole	Using exaggeration not meant to be taken literally	The girl weighed less than a feather.

Figurative language can be used to give additional insight into the theme or message of a text by moving beyond the usual and literal meaning of words and phrases. It can also be used to appeal to the senses of readers and create a more in-depth story.

Understanding the Development of Themes

Identifying Theme or Central Message
The *theme* is the central message of a fictional work, whether that work is structured as prose, drama, or poetry. It is the heart of what an author is trying to say to readers through the writing, and theme is largely conveyed through literary elements and techniques.

In literature, a theme can be often be determined by considering the over-arching narrative conflict with the work. Though there are several types of conflicts and several potential themes within them, the following are the most common:

- *Individual against the self*—relevant to themes of self-awareness, internal struggles, pride, coming of age, facing reality, fate, free will, vanity, loss of innocence, loneliness, isolation, fulfillment, failure, and disillusionment

- *Individual against nature*— relevant to themes of knowledge vs. ignorance, nature as beauty, quest for discovery, self-preservation, chaos and order, circle of life, death, and destruction of beauty

- *Individual against society*— relevant to themes of power, beauty, good, evil, war, class struggle, totalitarianism, role of men/women, wealth, corruption, change vs. tradition, capitalism, destruction, heroism, injustice, and racism

- *Individual against another individual*— relevant to themes of hope, loss of love or hope, sacrifice, power, revenge, betrayal, and honor

For example, in Hawthorne's *The Scarlet Letter*, one possible narrative conflict could be the individual against the self, with a relevant theme of internal struggles. This theme is alluded to through characterization—Dimmesdale's moral struggle with his love for Hester and Hester's internal struggles with the truth and her daughter, Pearl. It's also alluded to through plot—Dimmesdale's suicide and Hester helping the very townspeople who initially condemned her.

Sometimes, a text can convey a *message* or *universal lesson*—a truth or insight that the reader infers from the text, based on analysis of the literary and/or poetic elements. This message is often presented as a statement. For example, a potential message in Shakespeare's *Hamlet* could be "Revenge is what ultimately drives the human soul." This message can be immediately determined through plot and characterization in numerous ways, but it can also be determined through the setting of Norway, which is bordering on war.

How Authors Develop Theme

Authors employ a variety of techniques to present a theme. They may compare or contrast characters, events, places, ideas, or historical or invented settings to speak thematically. They may use analogies, metaphors, similes, allusions, or other literary devices to convey the theme. An author's use of diction, syntax, and tone can also help convey the theme. Authors will often develop themes through the development of characters, use of the setting, repetition of ideas, use of symbols, and through contrasting value systems. Authors of both fiction and nonfiction genres will use a variety of these techniques to develop one or more themes.

Regardless of the literary genre, there are commonalities in how authors, playwrights, and poets develop themes or central ideas.

Authors often do research, the results of which contributes to theme. In prose fiction and drama, this research may include real historical information about the setting the author has chosen or include elements that make fictional characters, settings, and plots seem realistic to the reader. In nonfiction, research is critical since the information contained within this literature must be accurate and, moreover, accurately represented.

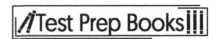

In fiction, authors present a narrative conflict that will contribute to the overall theme. In fiction, this conflict may involve the storyline itself and some trouble within characters that needs resolution. In nonfiction, this conflict may be an explanation or commentary on factual people and events.

Authors will sometimes use character motivation to convey theme, such as in the example from *Hamlet* regarding revenge. In fiction, the characters an author creates will think, speak, and act in ways that effectively convey the theme to readers. In nonfiction, the characters are factual, as in a biography, but authors pay particular attention to presenting those motivations to make them clear to readers.

Authors also use literary devices as a means of conveying theme. For example, the use of moon symbolism in Mary Shelley's *Frankenstein* is significant as its phases can be compared to the phases that the Creature undergoes as he struggles with his identity.

The selected point of view can also contribute to a work's theme. The use of first-person point of view in a fiction or non-fiction work engages the reader's response differently than third person point of view. The central idea or theme from a first person narrative may differ from a third-person limited text.

In literary nonfiction, authors usually identify the purpose of their writing, which differs from fiction, where the general purpose is to entertain. The purpose of nonfiction is usually to inform, persuade, or entertain the audience. The stated purpose of a non-fiction text will drive how the central message or theme, if applicable, is presented.

Authors identify an audience for their writing, which is critical in shaping the theme of the work. For example, the audience for J.K. Rowling's *Harry Potter* series would be different than the audience for a biography of George Washington. The audience an author chooses to address is closely tied to the purpose of the work. The choice of an audience also drives the choice of language and level of diction an author uses. Ultimately, the intended audience determines the level to which that subject matter is presented and the complexity of the theme.

Point of View

As mentioned, point of view is an important writing device to consider. In fiction writing, point of view refers to who tells the story or from whose perspective readers are observing as they read. In non-fiction writing, the *point of view* refers to whether the author refers to himself/herself, his/her readers, or chooses not to mention either. Whether fiction or nonfiction, the author will carefully consider the impact the perspective will have on the purpose and main point of the writing.

- *First-person point of view*: The story is told from the writer's perspective. In fiction, this would mean that the main character is also the narrator. First-person point of view is easily recognized by the use of personal pronouns such as *I, me, we, us, our, my,* and *myself*.

- *Third-person point of view*: In a more formal essay, this would be an appropriate perspective because the focus should be on the subject matter, not the writer or the reader. Third-person point of view is recognized by the use of the pronouns *he, she, they,* and *it*. In fiction writing, third person point of view has a few variations.

- *Third-person limited* point of view refers to a story told by a narrator who has access to the thoughts and feelings of just one character.

- In *third-person omniscient* point of view, the narrator has access to the thoughts and feelings of all the characters.

- In *third-person objective* point of view, the narrator is like a fly on the wall and can see and hear what the characters do and say but does not have access to their thoughts and feelings.

- *Second-person point of view*: This point of view isn't commonly used in fiction or non-fiction writing because it directly addresses the reader using the pronouns *you*, *your*, and *yourself*. Second-person perspective is more appropriate in direct communication, such as business letters or emails.

Point of View	Pronouns Used
First person	I, me, we, us, our, my, myself
Second person	You, your, yourself
Third person	He, she, it, they

Style, Tone, and Mood

Style, tone, and mood are often thought to be the same thing. Though they're closely related, there are important differences to keep in mind. The easiest way to do this is to remember that style "creates and affects" tone and mood. More specifically, style is how the writer uses words to create the desired tone and mood for their writing.

Style

Style can include any number of technical writing choices. A few examples of style choices include:

- Sentence Construction: When presenting facts, does the writer use shorter sentences to create a quicker sense of the supporting evidence, or do they use longer sentences to elaborate and explain the information?

- Technical Language: Does the writer use jargon to demonstrate their expertise in the subject, or do they use ordinary language to help the reader understand things in simple terms?

- Formal Language: Does the writer refrain from using contractions such as *won't* or *can't* to create a more formal tone, or do they use a colloquial, conversational style to connect to the reader?

- Formatting: Does the writer use a series of shorter paragraphs to help the reader follow a line of argument, or do they use longer paragraphs to examine an issue in great detail and demonstrate their knowledge of the topic?

On the test, examine the writer's style and how their writing choices affect the way the text comes across.

Tone

Tone refers to the writer's attitude toward the subject matter. Tone is usually explained in terms of a work of fiction. For example, the tone conveys how the writer feels about their characters and the situations in which they're involved. Nonfiction writing is sometimes thought to have no tone at all; however, this is incorrect.

A lot of nonfiction writing has a neutral tone, which is an important tone for the writer to take. A neutral tone demonstrates that the writer is presenting a topic impartially and letting the information speak for itself. On the other hand, nonfiction writing can be just as effective and appropriate if the tone isn't neutral. For instance, take this example involving seat belts:

> Seat belts save more lives than any other automobile safety feature. Many studies show that airbags save lives as well; however, not all cars have airbags. For instance, some older cars don't. Furthermore, air bags aren't entirely reliable. For example, studies show that in 15% of accidents airbags don't deploy as designed, but, on the other hand, seat belt malfunctions are extremely rare. The number of highway fatalities has plummeted since laws requiring seat belt usage were enacted.

In this passage, the writer mostly chooses to retain a neutral tone when presenting information. If the writer would instead include their own personal experience of losing a friend or family member in a car accident, the tone would change dramatically. The tone would no longer be neutral and would show that the writer has a personal stake in the content, allowing them to interpret the information in a different way. When analyzing tone, consider what the writer is trying to achieve in the text and how they *create* the tone using style.

Mood
Mood refers to the feelings and atmosphere that the writer's words create for the reader. Like tone, many nonfiction texts can have a neutral mood. To return to the previous example, if the writer would choose to include information about a person they know being killed in a car accident, the text would suddenly carry an emotional component that is absent in the previous example. Depending on how they present the information, the writer can create a sad, angry, or even hopeful mood. When analyzing the mood, consider what the writer wants to accomplish and whether the best choice was made to achieve that end.

Drawing Conclusions

When drawing conclusions about texts or passages, readers should do two main things: 1) Use the information that they already know and 2) Use the information they have learned from the text or passage. Authors write with an intended purpose, and it is the readers' responsibility to understand and form logical conclusions of authors' ideas. It is important to remember that the readers' conclusions should be supported by information directly from the text. Readers cannot simply form conclusions based off of only information they already know.

There are several ways readers can draw conclusions from authors' ideas and points to consider when doing so, such as text evidence, text credibility, and directly stated information versus implications.

Text Evidence
Text evidence is the information readers find in a text or passage that supports the main idea or point(s) in a story. In turn, text evidence can help readers draw conclusions about the text or passage. The information should be taken directly from the text or passage and placed in quotation marks. Text evidence provides readers with information to support ideas about the text or passage so that they simply do not just rely on their own thoughts. Details should be precise, descriptive, and factual. Statistics are a great piece of text evidencebecause it provides readers with exact numbers and not just a generalization. For example, instead of saying "Asia has a larger population than Europe," authors could provide detailed information such as "In Asia there are over 7 billion people, whereas in Europe

there are a little over 750 million." More definitive information provides better evidence to readers to help support their conclusions about texts or passages.

Text Credibility

Credible sources are important when drawing conclusions because readers need to be able to trust what they are reading. Authors should always use credible sources to help gain the trust of their readers. A text is *credible* when it is believable and the author is objective and unbiased. If readers do not trust authors' words, they may simply dismiss the text completely. For example, if an author writes a persuasive essay, he or she is outwardly trying to sway readers' opinions to align with his or her own, providing readers with the liberty to do what they please with the text. Readers may agree or disagree with the author, which may, in turn, lead them to believe that the author is credible or not credible. Also, readers should keep in mind the source of the text. If readers review a journal about astronomy, would a more reliable source be a NASA employee or a plumber? Overall, text credibility is important when drawing conclusions because readers want reliable sources that support the decisions they have made about the author's ideas.

Directly Stated Information Versus Implications

Engaged readers should constantly self-question while reviewing texts to help them form conclusions. Self-questioning is when readers review a paragraph, page, passage, or chapter and ask themselves, "Did I understand what I read?," "What was the main event in this section?," "Where is this taking place?," and so on. Authors can provide clues or pieces of evidence throughout a text or passage to guide readers toward a conclusion. This is why active and engaged readers should read the text or passage in its entirety before forming a definitive conclusion. If readers do not gather all the necessary pieces of evidence, then they may jump to an illogical conclusion.

At times, authors directly state conclusions while others simply imply them. Of course, it is easier if authors outwardly provide conclusions to readers because this does not leave any information open to interpretation. However, implications are things that authors do not directly state but can be assumed based off of information they provided. If authors only imply what may have happened, readers can form a menagerie of ideas for conclusions. For example, in the statement: *Once we heard the sirens, we hunkered down in the storm shelter*, the author does not directly state that there was a tornado, but clues such as "sirens" and "storm shelter" provide insight to the readers to help form that conclusion.

Opinions, Facts, and Fallacies

As mentioned previously, authors write with a purpose. They adjust their writing for an intended audience. It is the readers' responsibility to comprehend the writing style or purpose of the author. When readers understand a writer's purpose, they can then form their own thoughts about the text(s) regardless of whether their thoughts are the same as or different from the author's. The following section will examine different writing tactics that authors use, such as facts versus opinions, bias and stereotypes, appealing to the readers' emotions, and fallacies (including false analogies, circular reasoning, false dichotomy, and overgeneralization).

Facts Versus Opinions

Readers need to be aware of the writer's purpose to help discern facts and opinions within texts. A *fact* is a piece of information that is true. It can either prove or disprove claims or arguments presented in texts. Facts cannot be changed or altered. For example, the statement: *Abraham Lincoln was assassinated on April 15, 1865*, is a fact. The date and related events cannot be altered.

Authors not only present facts in their writing to support or disprove their claim(s), but they may also express their opinions. Authors may use facts to support their own opinions, especially in a persuasive text; however, that does not make their opinions facts. An *opinion* is a belief or view formed about something that is not necessarily based on the truth. Opinions often express authors' personal feelings about a subject and use words like *believe, think,* or *feel.* For example, the statement: *Abraham Lincoln was the best president who has ever lived*, expresses the writer's opinion. Not all writers or readers agree or disagree with the statement. Therefore, the statement can be altered or adjusted to express opposing or supporting beliefs, such as "Abraham Lincoln was the worst president who has ever lived" or "I also think Abraham Lincoln was a great president."

When authors include facts and opinions in their writing, readers may be less influenced by the text(s). Readers need to be conscious of the distinction between facts and opinions while going through texts. Not only should the intended audience be vigilant in following authors' thoughts versus valid information, readers need to check the source of the facts presented. Facts should have reliable sources derived from credible outlets like almanacs, encyclopedias, medical journals, and so on.

Bias and Stereotypes
Not only can authors state facts or opinions in their writing, they sometimes intentionally or unintentionally show bias or portray a stereotype. A *bias* is when someone demonstrates a prejudice in favor of or against something or someone in an unfair manner. When an author is biased in his or her writing, readers should be skeptical despite the fact that the author's bias may be correct. For example, two athletes competed for the same position. One athlete is related to the coach and is a mediocre athlete, while the other player excels and deserves the position. The coach chose the less talented player who is related to him for the position. This is a biased decision because it favors someone in an unfair way.

Similar to a bias, a *stereotype* shows favoritism or opposition but toward a specific group or place. Stereotypes create an oversimplified or overgeneralized idea about a certain group, person, or place. For example,

> Women are horrible drivers.

This statement basically labels *all* women as horrible drivers. While there may be some terrible female drivers, the stereotype implies that *all* women are bad drivers when, in fact, not *all* women are. While many readers are aware of several vile ethnic, religious, and cultural stereotypes, audiences should be cautious of authors' flawed assumptions because they can be less obvious than the despicable examples that are unfortunately pervasive in society.

Identifying Rhetorical Strategies

Rhetoric refers to an author's use of particular strategies, appeals, and devices to persuade an intended audience. The more effective the use of rhetoric, the more likely the audience will be persuaded.

Determining an Author's Point of View
A *rhetorical strategy*—also referred to as a *rhetorical mode*—is the structural way an author chooses to present his/her argument. Though the terms noted below are similar to the organizational structures noted earlier, these strategies do not imply that the entire text follows the approach. For example, a cause and effect organizational structure is solely that, nothing more. A persuasive text may use cause and effect as a strategy to convey a singular point. Thus, an argument may include several of the

strategies as the author strives to convince his or her audience to take action or accept a different point of view. It's important that readers are able to identify an author's thesis and position on the topic in order to be able to identify the careful construction through which the author speaks to the reader. The following are some of the more common rhetorical strategies:

- *Cause and effect*—establishing a logical correlation or causation between two ideas

- *Classification/division*—the grouping of similar items together or division of something into parts

- *Comparison/contrast*—the distinguishing of similarities/differences to expand on an idea

- *Definition*—used to clarify abstract ideas, unfamiliar concepts, or to distinguish one idea from another

- *Description*—use of vivid imagery, active verbs, and clear adjectives to explain ideas

- *Exemplification*—the use of examples to explain an idea

- *Narration*—anecdotes or personal experience to present or expand on a concept

- *Problem/Solution*—presentation of a problem or problems, followed by proposed solution(s)

Rhetorical Strategies and Devices

A *rhetorical device* is the phrasing and presentation of an idea that reinforces and emphasizes a point in an argument. A rhetorical device is often quite memorable. One of the more famous uses of a rhetorical device is in John F. Kennedy's 1961 inaugural address: "Ask not what your country can do for you, ask what you can do for your country." The contrast of ideas presented in the phrasing is an example of the rhetorical device of antimetabole. Some other common examples are provided below, but test takers should be aware that this is not a complete list.

Device	Definition	Example
Allusion	A reference to a famous person, event, or significant literary text as a form of significant comparison	"We are apt to shut our eyes against a painful truth, and listen to the song of that siren till she transforms us into beasts." Patrick Henry
Anaphora	The repetition of the same words at the beginning of successive words, phrases, or clauses, designed to emphasize an idea	"We shall not flag or fail. We shall go on to the end. We shall fight in France, we shall fight on the seas and oceans, we shall fight with growing confidence … we shall fight in the fields and in the streets, we shall fight in the hills. We shall never surrender." Winston Churchill
Understatement	A statement meant to portray a situation as less important than it actually is to create an ironic effect	"The war in the Pacific has not necessarily developed in Japan's favor." Emperor Hirohito, surrendering Japan in World War II
Parallelism	A syntactical similarity in a structure or series of structures used for impact of an idea, making it memorable	"A penny saved is a penny earned." Ben Franklin
Rhetorical question	A question posed that is not answered by the writer though there is a desired response, most often designed to emphasize a point	"Can anyone look at our reduced standing in the world today and say, 'Let's have four more years of this?'" Ronald Reagan

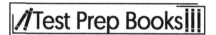

Understanding Methods Used to Appeal to a Specific Audience

Rhetorical Appeals

In an argument or persuasive text, an author will strive to sway readers to an opinion or conclusion. To be effective, an author must consider his or her intended audience. Although an author may write text for a general audience, he or she will use methods of appeal or persuasion to convince that audience. Aristotle asserted that there were three methods or modes by which a person could be persuaded. These are referred to as *rhetorical appeals*.

The three main types of rhetorical appeals are shown in the following graphic.

Ethos, also referred to as an *ethical appeal*, is an appeal to the audience's perception of the writer as credible (or not), based on their examination of their ethics and who the writer is, his/her experience or incorporation of relevant information, or his/her argument. For example, authors may present testimonials to bolster their arguments. The reader who critically examines the veracity of the testimonials and the credibility of those giving the testimony will be able to determine if the author's use of testimony is valid to his or her argument. In turn, this will help the reader determine if the author's thesis is valid. An author's careful and appropriate use of technical language can create an overall knowledgeable effect and, in turn, act as a convincing vehicle when it comes to credibility. Overuse of technical language, however, may create confusion in readers and obscure an author's overall intent.

Pathos, also referred to as a *pathetic* or *emotional appeal*, is an appeal to the audience's sense of identity, self-interest, or emotions. A critical reader will notice when the author is appealing to pathos through anecdotes and descriptions that elicit an emotion such as anger or pity. Readers should also beware of factual information that uses generalization to appeal to the emotions. While it's tempting to believe an author is the source of truth in his or her text, an author who presents factual information as universally true, consistent throughout time, and common to all groups is using *generalization*. Authors

who exclusively use generalizations without specific facts and credible sourcing are attempting to sway readers solely through emotion.

Logos, also referred to as a *logical appeal*, is an appeal to the audience's ability to see and understand the logic in a claim offered by the writer. A critical reader has to be able to evaluate an author's arguments for validity of reasoning and for sufficiency when it comes to argument.

Understanding the Effect of Word Choice

An author's choice of words—also referred to as *diction*—helps to convey his or her meaning in a particular way. Through diction, an author can convey a particular tone—e.g., a humorous tone, a serious tone—in order to support the thesis in a meaningful way to the reader.

Connotation and Denotation

Connotation is when an author chooses words or phrases that invoke ideas or feelings other than their literal meaning. An example of the use of connotation is the word *cheap*, which suggests something is poor in value or negatively describes a person as reluctant to spend money. When something or someone is described this way, the reader is more inclined to have a particular image or feeling about it or him/her. Thus, connotation can be a very effective language tool in creating emotion and swaying opinion. However, connotations are sometimes hard to pin down because varying emotions can be associated with a word. Generally, though, connotative meanings tend to be fairly consistent within a specific cultural group.

Denotation refers to words or phrases that mean exactly what they say. It is helpful when a writer wants to present hard facts or vocabulary terms with which readers may be unfamiliar. Some examples of denotation are the words *inexpensive* and *frugal*. *Inexpensive* refers to the cost of something, not its value, and *frugal* indicates that a person is conscientiously watching his or her spending. These terms do not elicit the same emotions that *cheap* does.

Authors sometimes choose to use both, but what they choose and when they use it is what critical readers need to differentiate. One method isn't inherently better than the other; however, one may create a better effect, depending upon an author's intent. If, for example, an author's purpose is to inform, to instruct, and to familiarize readers with a difficult subject, his or her use of connotation may be helpful. However, it may also undermine credibility and confuse readers. An author who wants to create a credible, scholarly effect in his or her text would most likely use denotation, which emphasizes literal, factual meaning and examples.

Technical Language

Test takers and critical readers alike should be very aware of technical language used within informational text. *Technical language* refers to terminology that is specific to a particular industry and is best understood by those specializing in that industry. This language is fairly easy to differentiate, since it will most likely be unfamiliar to readers. It's critical to be able to define technical language either by the author's written definition, through the use of an included glossary—if offered—or through context clues that help readers clarify word meaning.

Identifying the Position and Purpose

When it comes to authors' writings, readers should always identify a position or stance. No matter how objective a piece may seem, assume the author has preconceived beliefs. Reduce the likelihood of accepting an invalid argument by looking for multiple articles on the topic, including those with varying opinions. If several opinions point in the same direction, and are backed by reputable peer-reviewed sources, it's more likely the author has a valid argument. Positions that run contrary to widely held beliefs and existing data should invite scrutiny. There are exceptions to the rule, so be a careful consumer of information.

Though themes, symbols, and motifs are buried deep within the text and can sometimes be difficult to infer, an author's purpose is usually obvious from the beginning. There are four purposes of writing: to inform, to persuade, to describe, and to entertain. Informative writings present facts in an accessible way. Persuasive writing appeals to emotions and logic to inspire the reader to adopt a specific stance. Be wary of this type of writing, as it often lacks objectivity. Descriptive writing is designed to paint a picture in the reader's mind, while writings that entertain are often narratives designed to engage and delight the reader.

The various writing styles are usually blended, with one purpose dominating the rest. For example, a persuasive piece might begin with a humorous tale to make readers more receptive to the persuasive message, or a recipe in a cookbook designed to inform might be preceded by an entertaining anecdote that makes the recipe more appealing.

Apply Information

A natural extension of being able to make an inference from a given set of information is also being able to apply that information to a new context. This is especially useful in non-fiction or informative writing. Considering the facts and details presented in the text, readers should consider how the same information might be relevant in a different situation. The following is an example of applying an inferential conclusion to a different context:

> Often, individuals behave differently in large groups than they do as individuals. One example of this is the psychological phenomenon known as the bystander effect. According to the bystander effect, the more people who witness an accident or crime occur, the less likely each individual bystander is to respond or offer assistance to the victim. A classic example of this is the murder of Kitty Genovese in New York City in the 1960s. Although there were over thirty witnesses to her killing by a stabber, none of them intervened to help Kitty or contact the police.

Considering the phenomenon of the bystander effect, what would probably happen if somebody tripped on the stairs in a crowded subway station?
a. Everybody would stop to help the person who tripped
b. Bystanders would point and laugh at the person who tripped
c. Someone would call the police after walking away from the station
d. Few if any bystanders would offer assistance to the person who tripped

This question asks readers to apply the information they learned from the passage, which is an informative paragraph about the bystander effect. According to the passage, this is a concept in psychology that describes the way people in groups respond to an accident—the more people are present, the less likely any one person is to intervene. While the passage illustrates this effect with the example of a woman's murder, the question asks readers to apply it to a different context—in this case, someone falling down the stairs in front of many subway passengers. Although this specific situation is not discussed in the passage, readers should be able to apply the general concepts described in the paragraph. The definition of the bystander effect includes any instance of an accident or crime in front

of a large group of people. The question asks about a situation that falls within the same definition, so the general concept should still hold true: in the midst of a large crowd, few individuals are likely to actually respond to an accident. In this case, answer choice (d) is the best response.

Inferences in a Text

Readers should be able to make *inferences*. Making an inference requires the reader to read between the lines and look for what is *implied* rather than what is directly stated. That is, using information that is known from the text, the reader is able to make a logical assumption about information that is *not* directly stated but is probably true. Read the following passage:

"Hey, do you wanna meet my new puppy?" Jonathan asked.

"Oh, I'm sorry but please don't—" Jacinta began to protest, but before she could finish, Jonathan had already opened the passenger side door of his car and a perfect white ball of fur came bouncing towards Jacinta.

"Isn't he the cutest?" beamed Jonathan.

"Yes—achoo!—he's pretty—aaaachooo!!—adora—aaa—aaaachoo!" Jacinta managed to say in between sneezes. "But if you don't mind, I—I—achoo!—need to go inside."

Which of the following can be inferred from Jacinta's reaction to the puppy?
a. she hates animals
b. she is allergic to dogs
c. she prefers cats to dogs
d. she is angry at Jonathan

An inference requires the reader to consider the information presented and then form their own idea about what is probably true. Based on the details in the passage, what is the best answer to the question? Important details to pay attention to include the tone of Jacinta's dialogue, which is overall polite and apologetic, as well as her reaction itself, which is a long string of sneezes. Answer choices (a) and (d) both express strong emotions ("hates" and "angry") that are not evident in Jacinta's speech or actions. Answer choice (c) mentions cats, but there is nothing in the passage to indicate Jacinta's feelings about cats. Answer choice (b), "she is allergic to dogs," is the most logical choice—based on the fact that she began sneezing as soon as a fluffy dog approached her, it makes sense to guess that Jacinta might be allergic to dogs. So even though Jacinta never directly states, "Sorry, I'm allergic to dogs!" using the clues in the passage, it is still reasonable to guess that this is true.

Making inferences is crucial for readers of literature because literary texts often avoid presenting complete and direct information to readers about characters' thoughts or feelings, or they present this information in an unclear way, leaving it up to the reader to interpret clues given in the text. In order to make inferences while reading, readers should ask themselves:

- What details are being presented in the text?
- Is there any important information that seems to be missing?
- Based on the information that the author *does* include, what else is probably true?
- Is this inference reasonable based on what is already known?

Understanding the Task, Purpose, and Audience

Identifying the Task, Purpose, and Intended Audience

An author's *writing style*—the way in which words, grammar, punctuation, and sentence fluidity are used—is the most influential element in a piece of writing, and it is dependent on the purpose and the audience for whom it is intended. Together, a writing style and mode of writing form the foundation of a written work, and a good writer will choose the most effective mode and style to convey a message to readers.

Writers should first determine what they are trying to say and then choose the most effective mode of writing to communicate that message. Different writing modes and *word choices* will affect the tone of a piece—that is, its underlying attitude, emotion, or character. The argumentative mode may utilize words that are earnest, angry, passionate, or excited whereas an informative piece may have a sterile, germane, or enthusiastic tone. The tones found in narratives vary greatly, depending on the purpose of the writing. *Tone* will also be affected by the audience—teaching science to children or those who may be uninterested would be most effective with enthusiastic language and exclamation points whereas teaching science to college students may take on a more serious and professional tone, with fewer charged words and punctuation choices that are inherent to academia.

Sentence fluidity—whether sentences are long and rhythmic or short and succinct—also affects a piece of writing as it determines the way in which a piece is read. Children or audiences unfamiliar with a subject do better with short, succinct sentence structures as these break difficult concepts up into shorter points. A period, question mark, or exclamation point is literally a signal for the reader to stop and takes more time to process. Thus, longer, more complex sentences are more appropriate for adults or educated audiences as they can fit more information in between processing time.

The amount of *supporting detail* provided is also tailored to the audience. A text that introduces a new subject to its readers will focus more on broad ideas without going into greater detail whereas a text that focuses on a more specific subject is likely to provide greater detail about the ideas discussed.

Writing styles, like modes, are most effective when tailored to their audiences. Having awareness of an audience's demographic is one of the most crucial aspects of properly communicating an argument, a story, or a set of information.

Choosing the Most Appropriate Type of Writing

Before beginning any writing, it is imperative that a writer have a firm grasp on the message he or she wishes to convey and how he or she wants readers to be affected by the writing. For example, does the author want readers to be more informed about the subject? Does the writer want readers to agree with his or her opinion? Does the writer want readers to get caught up in an exciting narrative? The following steps are a guide to determining the appropriate type of writing for a task, purpose, and audience:

1. Identifying the purpose for writing the piece
2. Determining the audience
3. Adapting the writing mode, word choices, tone, and style to fit the audience and the purpose

It is important to distinguish between a work's purpose and its main idea. The essential difference between the two is that the *main idea* is what the author wants to communicate about the topic at hand whereas the *primary purpose* is why the author is writing in the first place. The primary purpose is

what will determine the type of writing an author will choose to utilize, not the main idea, though the two are related. For example, if an author writes an article on the mistreatment of animals in factory farms and, at the end, suggests that people should convert to vegetarianism, the main idea is that vegetarianism would reduce the poor treatment of animals. The primary purpose is to convince the reader to stop eating animals. Since the primary purpose is to galvanize an audience into action, the author would choose the argumentative writing mode.

The next step is to consider to whom the author is appealing as this will determine the type of details to be included, the diction to be used, the tone to be employed, and the sentence structure to be used. An audience can be identified by considering the following questions:

- What is the purpose for writing the piece?

- To whom is it being written?

- What is their age range?

- Are they familiar with the material being presented, or are they just being newly introduced to it?

- Where are they from?

- Is the task at hand in a professional or casual setting?

- Is the task at hand for monetary gain?

These are just a few of the numerous considerations to keep in mind, but the main idea is to become as familiar with the audience as possible. Once the audience has been understood, the author can then adapt the writing style to align with the readers' education and interests. The audience is what determines the *rhetorical appeal* the author will use—ethos, pathos, or logos. *Ethos* is a rhetorical appeal to an audience's ethics and/or morals. Ethos is most often used in argumentative and informative writing modes. *Pathos* is an appeal to the audience's emotions and sympathies, and it is found in argumentative, descriptive, and narrative writing modes. *Logos* is an appeal to the audience's logic and reason and is used primarily in informative texts as well as in supporting details for argumentative pieces. Rhetorical appeals are discussed in depth in the informational texts and rhetoric section of the test.

If the author is trying to encourage global conversion to vegetarianism, he or she may choose to use all three rhetorical appeals to reach varying personality types. Those who are less interested in the welfare of animals but are interested in facts and science would relate more to logos. Animal lovers would relate better to an emotional appeal. In general, the most effective works utilize all three appeals.

Finally, after determining the writing mode and rhetorical appeal, the author will consider word choice, sentence structure, and tone, depending on the purpose and audience. The author may choose words that convey sadness or anger when speaking about animal welfare if writing to persuade, or he or she will stick to dispassionate and matter-of-fact tones, if informing the public on the treatment of animals in factory farms. If the author is writing to a younger or less-educated audience, he or she may choose to shorten and simplify sentence structures and word choice. If appealing to an audience with more expert knowledge on a particular subject, writers will more likely employ a style of longer sentences and more complex vocabulary.

Depending on the task, the author may choose to use a first person, second person, or third person point of view. First person and second person perspectives are inherently more casual in tone, including the author and the reader in the rhetoric, while third person perspectives are often seen in more professional settings.

Evaluating the Effectiveness of a Piece of Writing

An effective and engaging piece of writing will cause the reader to forget about the author entirely. Readers will become so engrossed in the subject, argument, or story at hand that they will almost identify with it, readily adopting beliefs proposed by the author or accepting all elements of the story as believable. On the contrary, poorly written works will cause the reader to be hyperaware of the author, doubting the writer's knowledge of a subject or questioning the validity of a narrative. Persuasive or expository works that are poorly researched will have this effect, as well as poorly planned stories with significant plot holes. An author must consider the task, purpose, and audience to sculpt a piece of writing effectively.

When evaluating the effectiveness of a piece, the most important thing to consider is how well the purpose is conveyed to the audience through the mode, use of rhetoric, and writing style.

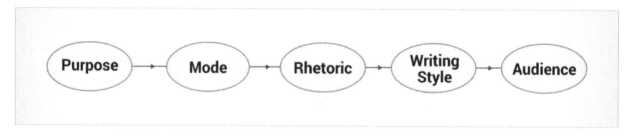

The purpose must pass through these three aspects for effective delivery to the audience. If any elements are not properly considered, the reader will be overly aware of the author, and the message will be lost. The following is a checklist for evaluating the effectiveness of a piece:

- Does the writer choose the appropriate writing mode—argumentative, narrative, descriptive, informative—for his or her purpose?

- Does the writing mode employed contain characteristics inherent to that mode?

- Does the writer consider the personalities/interests/demographics of the intended audience when choosing rhetorical appeals?

- Does the writer use appropriate vocabulary, sentence structure, voice, and tone for the audience demographic?

- Does the author properly establish himself/herself as having authority on the subject, if applicable?

- Does the piece make sense?

Another thing to consider is the medium in which the piece was written. If the medium is a blog, diary, or personal letter, the author may adopt a more casual stance towards the audience. If the piece of writing is a story in a book, a business letter or report, or a published article in a journal or if the task is to gain money or support or to get published, the author may adopt a more formal stance. Ultimately, the writer will want to be very careful in how he or she addresses the reader.

Finally, the effectiveness of a piece can be evaluated by asking how well the purpose was achieved. For example, if students are assigned to read a persuasive essay, instructors can ask whether the author influences students' opinions. Students may be assigned two differing persuasive texts with opposing perspectives and be asked which writer was more convincing. Students can then evaluate what factors contributed to this—for example, whether one author uses more credible supporting facts, appeals more effectively to readers' emotions, presents more believable personal anecdotes, or offers stronger counterargument refutation. Students can then use these evaluations to strengthen their own writing skills.

Text Structure

Depending on what the author is attempting to accomplish, certain formats or text structures work better than others. For example, a sequence structure might work for narration but not when identifying similarities and differences between dissimilar concepts. Similarly, a comparison-contrast structure is not useful for narration. It's the author's job to put the right information in the correct format.

Readers should be familiar with the five main literary structures:

1. *Sequence* structure (sometimes referred to as the order structure) is when the order of events proceed in a predictable order. In many cases, this means the text goes through the plot elements: exposition, rising action, climax, falling action, and resolution. Readers are introduced to characters, setting, and conflict in the exposition. In the rising action, there's an increase in tension and suspense. The climax is the height of tension and the point of no return. Tension decreases during the falling action. In the resolution, any conflicts presented in the exposition are solved, and the story concludes. An informative text that is structured sequentially will often go in order from one step to the next.

2. In the *problem-solution* structure, authors identify a potential problem and suggest a solution. This form of writing is usually divided into two paragraphs and can be found in informational texts. For example, cell phone, cable, and satellite providers use this structure in manuals to help customers troubleshoot or identify problems with services or products.

3. When authors want to discuss similarities and differences between separate concepts, they arrange thoughts in a *comparison-contrast* paragraph structure. Venn diagrams are an effective graphic organizer for comparison-contrast structures because they feature two overlapping circles that can be used to organize similarities and differences. A comparison-contrast essay organizes one paragraph based on similarities and another based on differences. A comparison-contrast essay can also be arranged with the similarities and differences of individual traits addressed within individual paragraphs. Words such as *however*, *but*, and *nevertheless* help signal a contrast in ideas.

4. *Descriptive* writing structure is designed to appeal to your senses. Much like an artist who constructs a painting, good descriptive writing builds an image in the reader's mind by appealing to the five senses: sight, hearing, taste, touch, and smell. However, overly descriptive writing can become tedious; sparse descriptions can make settings and characters seem flat. Good authors strike a balance by applying descriptions only to passages, characters, and settings that are integral to the plot.

5. Passages that use the *cause and effect* structure are simply asking *why* by demonstrating some type of connection between ideas. Words such as *if*, *since*, *because*, *then*, or *consequently* indicate relationship. By switching the order of a complex sentence, the writer can rearrange the emphasis on different clauses. Saying *"If Sheryl is late, we'll miss the dance"* is different from saying "We'll miss the

dance if Sheryl is late." One emphasizes Sheryl's tardiness while the other emphasizes missing the dance. Paragraphs can also be arranged in a cause and effect format. Since the format—before and after—is sequential, it is useful when authors wish to discuss the impact of choices. Researchers often apply this paragraph structure to the scientific method.

Practice Questions

Questions 1-5 are based on the following passage:

When researchers and engineers undertake a large-scale scientific project, they may end up making discoveries and developing technologies that have far wider uses than originally intended. This is especially true in NASA, one of the most influential and innovative scientific organizations in America. NASA spinoff technology refers to innovations originally developed for NASA space projects that are now used in a wide range of different commercial fields. Many consumers are unaware that products they are buying are based on NASA research! Spinoff technology proves that it is worthwhile to invest in science research because it could enrich people's lives in unexpected ways.

The first spinoff technology worth mentioning is baby food. In space, where astronauts have limited access to fresh food and fewer options about their daily meals, malnutrition is a serious concern. Consequently, NASA researchers were looking for ways to enhance the nutritional value of astronauts' food. Scientists found that a certain type of algae could be added to food, improving the food's neurological benefits. When experts in the commercial food industry learned of this algae's potential to boost brain health, they were quick to begin their own research. The nutritional substance from algae then developed into a product called life's DHA, which can be found in over 90% of infant food sold in America.

Another intriguing example of a spinoff technology can be found in fashion. People who are always dropping their sunglasses may have invested in a pair of sunglasses with scratch resistant lenses—that is, it's impossible to scratch the glass, even if the glasses are dropped on an abrasive surface. This innovation is incredibly advantageous for people who are clumsy, but most shoppers don't know that this technology was originally developed by NASA. Scientists first created scratch resistant glass to help protect costly and crucial equipment from getting scratched in space, especially the helmet visors in space suits. However, sunglasses companies later realized that this technology could be profitable for their products, and they licensed the technology from NASA.

1. What is the main purpose of this article?
   a. To advise consumers to do more research before making a purchase
   b. To persuade readers to support NASA research
   c. To tell a narrative about the history of space technology
   d. To define and describe instances of spinoff technology

2. What is the organizational structure of this article?
   a. A general definition followed by more specific examples
   b. A general opinion followed by supporting arguments
   c. An important moment in history followed by chronological details
   d. A popular misconception followed by counterevidence

3. Why did NASA scientists research algae?
    a. They already knew algae was healthy for babies.
    b. They were interested in how to grow food in space.
    c. They were looking for ways to add health benefits to food.
    d. They hoped to use it to protect expensive research equipment.

4. Why does the author mention space suit helmets?
    a. To give an example of astronaut fashion
    b. To explain where sunglasses got their shape
    c. To explain how astronauts protect their eyes
    d. To give an example of valuable space equipment

5. Which statement would the author probably NOT agree with?
    a. Consumers don't always know the history of the products they are buying.
    b. Sometimes new innovations have unexpected applications.
    c. It is difficult to make money from scientific research.
    d. Space equipment is often very expensive.

Answer Explanations

1. D: To define and describe instances of spinoff technology. This is an example of a purpose question—*why* did the author write this? The article contains facts, definitions, and other objective information without telling a story or arguing an opinion. In this case, the purpose of the article is to inform the reader. The only answer choice that is related to giving information is answer Choice *D*: to define and describe.

2. A: A general definition followed by more specific examples. This organization question asks readers to analyze the structure of the essay. The topic of the essay is about spinoff technology; the first paragraph gives a general definition of the concept, while the following two paragraphs offer more detailed examples to help illustrate this idea.

3. C: They were looking for ways to add health benefits to food. This reading comprehension question can be answered based on the second paragraph—scientists were concerned about astronauts' nutrition and began researching useful nutritional supplements. A in particular is not true because it reverses the order of discovery (first NASA identified algae for astronaut use, and then it was further developed for use in baby food).

4. D: To give an example of valuable space equipment. This purpose question requires readers to understand the relevance of the given detail. In this case, the author mentions "costly and crucial equipment" before mentioning space suit visors, which are given as an example of something that is very valuable. *A* is not correct because fashion is only related to sunglasses, not to NASA equipment. *B* can be eliminated because it is simply not mentioned in the passage. While *C* seems like it could be a true statement, it is also not relevant to what is being explained by the author.

5. C: It is difficult to make money from scientific research. The article gives several examples of how businesses have been able to capitalize on NASA research, so it is unlikely that the author would agree with this statement. Evidence for the other answer choices can be found in the article: *A*, the author mentions that "many consumers are unaware that products they are buying are based on NASA research"; *B* is a general definition of spinoff technology; and *D* is mentioned in the final paragraph.

Written Expression

Understanding the Conventions of Standard English

Parts of Speech

The English language has eight parts of speech, each serving a different grammatical function.

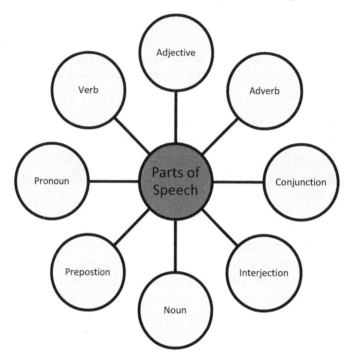

Verb

Verbs describe an action—e.g., *run*, *play*, *eat*—or a state of being—e.g., *is*, *are*, *was*. It is impossible to make a grammatically-complete sentence without a verb.

> He *runs* to the store.

> She *is* eight years old.

Noun

Nouns can be a person, place, or thing. They can refer to concrete objects—e.g., chair, apple, house—or abstract things—love, knowledge, friendliness.

> Look at the *dog*!

> Where are my *keys*?

Some nouns are *countable*, meaning they can be counted as separate entities—one chair, two chairs, three chairs. They can be either singular or plural. Other nouns, usually substances or concepts, are

uncountable—e.g., air, information, wealth—and some nouns can be both countable and uncountable depending on how they are used.

> I bought three *dresses*.

> *Respect* is important to me.

> I ate way too much *food* last night.

> At the international festival, you can sample *foods* from around the world.

Proper nouns are the specific names of people, places, or things and are almost always capitalized.

> <u>Marie Curie</u> studied at the <u>Flying University</u> in <u>Warsaw, Poland</u>.

Pronoun

Pronouns function as substitutes for nouns or noun phrases. Pronouns are often used to avoid constant repetition of a noun or to simplify sentences. *Personal pronouns* are used for people. Some pronouns are *subject pronouns*; they are used to replace the subject in a sentence—I, we, he, she, they.

> Is *he* your friend?

> *We* work together.

Object pronouns can function as the object of a sentence—me, us, him, her, them.

> Give the documents to *her*.

> Did you call *him* back yet?

Some pronouns can function as either the subject or the object—e.g., you, it. The subject of a sentence is the noun of the sentence that is doing or being something.

> *You* should try it.

> *It* tastes great.

Possessive pronouns indicate ownership. They can be used alone—mine, yours, his, hers, theirs, ours—or with a noun—my, your, his, her, their, ours. In the latter case, they function as a determiner, which is described in detail in the below section on adjectives.

> This table is *ours*.

> I can't find *my* phone!

Reflexive pronouns refer back to the person being spoken or written about. These pronouns end in -self/-selves.

> I've heard that New York City is gorgeous in the autumn, but I've never seen it for *myself*.

> After moving away from home, young people have to take care of *themselves*.

Indefinite pronouns are used for things that are unknown or unspecified. Some examples are *anybody, something,* and *everything.*

I'm looking for *someone* who knows how to fix computers.

I wanted to buy some shoes today, but I couldn't find *any* that I liked.

Adjective

An adjective modifies a noun, making it more precise or giving more information about it. Adjectives answer these questions: What kind? Which one?

I just bought a *red* car.

I don't like *cold* weather.

One special type of word that modifies a noun is a *determiner.* In fact, some grammarians classify determiners as a separate part of speech because whereas adjectives simply describe additional qualities of a noun, a determiner is often a necessary part of a noun phrase, without which the phrase is grammatically incomplete. A determiner indicates whether a noun is definite or indefinite, and can identify which noun is being discussed. It also introduces context to the noun in terms of quantity and possession. The most commonly-used determiners are articles—a, an, the.

I ordered *a* pizza.

She lives in *the* city.

Possessive pronouns discussed above, such as *my, your,* and *our,* are also determiners, along with *demonstratives*—this, that—and *quantifiers*—much, many, some. These determiners can take the place of an article.

Are you using *this* chair?

I need *some* coffee!

Adverb

Adverbs modify verbs, adjectives, and other adverbs. Words that end in –ly are usually adverbs. Adverbs answer these questions: When? Where? In what manner? To what degree?

She talks *quickly.*

The mountains are *incredibly* beautiful!

The students arrived *early.*

Please take your phone call *outside.*

Preposition

Prepositions show the relationship between different elements in a phrase or sentence and connect nouns or pronouns to other words in the sentence. Some examples of prepositions are words such as *after, at, behind, by, during, from, in, on, to,* and *with.*

> Let's go *to* class.

> Starry Night was painted *by* Vincent van Gogh *in* 1889.

Conjunction

Conjunctions join words, phrases, clauses, or sentences together, indicating the type of connection between these elements.

> I like pizza, *and* I enjoy spaghetti.

> I like to play baseball, *but* I'm allergic to mitts.

Some conjunctions are *coordinating*, meaning they give equal emphasis to two main clauses. Coordinating conjunctions are short, simple words that can be remembered using the mnemonic FANBOYS: for, and, nor, but, or, yet, so. Other conjunctions are *subordinating*. Subordinating conjunctions introduce dependent clauses and include words such as *because, since, before, after, if,* and *while.*

Interjection

An *interjection* is a short word that shows greeting or emotion. Examples of interjections include *wow, ouch, hey, oops, alas,* and *hey.*

> *Wow!* Look at that sunset!

> Was it your birthday yesterday? *Oops!* I forgot.

Errors in Standard English Grammar, Usage, Syntax, and Mechanics

Sentence Fragments

A *complete sentence* requires a verb and a subject that expresses a complete thought. Sometimes, the subject is omitted in the case of the implied *you*, used in sentences that are the command or imperative form—e.g., "Look!" or "Give me that." It is understood that the subject of the command is *you*, the listener or reader, so it is possible to have a structure without an explicit subject. Without these elements, though, the sentence is incomplete—it is a *sentence fragment*. While sentence fragments often occur in conversational English or creative writing, they are generally not appropriate in academic writing. Sentence fragments often occur when dependent clauses are not joined to an independent clause:

> *Sentence fragment*: Because the airline overbooked the flight.

The sentence above is a dependent clause that does not express a complete thought. What happened as a result of this cause? With the addition of an independent clause, this now becomes a complete sentence:

> *Complete sentence*: Because the airline overbooked the flight, several passengers were unable to board.

Sentences fragments may also occur through improper use of conjunctions:

> I'm going to the Bahamas for spring break. And to New York City for New Year's Eve.

While the first sentence above is a complete sentence, the second one is not because it is a prepositional phrase that lacks a subject [I] and a verb [am going]. Joining the two together with the coordinating conjunction forms one grammatically-correct sentence:

> I'm going to the Bahamas for spring break and to New York City for New Year's Eve.

Run-ons

A *run-on* is a sentence with too many independent clauses that are improperly connected to each other:

> This winter has been very cold some farmers have suffered damage to their crops.

The sentence above has two subject-verb combinations. The first is "this winter has been"; the second is "some farmers have suffered." However, they are simply stuck next to each other without any punctuation or conjunction. Therefore, the sentence is a run-on.

Another type of run-on occurs when writers use inappropriate punctuation:

> This winter has been very cold, some farmers have suffered damage to their crops.

Though a comma has been added, this sentence is still not correct. When a comma alone is used to join two independent clauses, it is known as a **comma splice**. Without an appropriate conjunction, a comma cannot join two independent clauses by itself.

Run-on sentences can be corrected by either dividing the independent clauses into two or more separate sentences or inserting appropriate conjunctions and/or punctuation. The run-on sentence can be amended by separating each subject-verb pair into its own sentence:

> This winter has been very cold. Some farmers have suffered damage to their crops.

The run-on can also be fixed by adding a comma and conjunction to join the two independent clauses with each other:

> This winter has been very cold, so some farmers have suffered damage to their crops.

Parallelism

Parallel structure occurs when phrases or clauses within a sentence contain the same structure. Parallelism increases readability and comprehensibility because it is easy to tell which sentence elements are paired with each other in meaning.

> Jennifer enjoys cooking, knitting, and to spend time with her cat.

This sentence is not parallel because the items in the list appear in two different forms. Some are *gerunds*, which is the verb + ing: *cooking, knitting*. The other item uses the *infinitive* form, which is to + verb: *to spend*. To create parallelism, all items in the list may reflect the same form:

> Jennifer enjoys cooking, knitting, and spending time with her cat.

All of the items in the list are now in gerund forms, so this sentence exhibits parallel structure. Here's another example:

> The company is looking for employees who are responsible and with a lot of experience.

Again, the items that are listed in this sentence are not parallel. "Responsible" is an adjective, yet "with a lot of experience" is a prepositional phrase. The sentence elements do not utilize parallel parts of speech.

> The company is looking for employees who are responsible and experienced.

"Responsible" and "experienced" are both adjectives, so this sentence now has parallel structure.

Dangling and Misplaced Modifiers

Modifiers enhance meaning by clarifying or giving greater detail about another part of a sentence. However, incorrectly-placed modifiers have the opposite effect and can cause confusion. A *misplaced modifier* is a modifier that is not located appropriately in relation to the word or phrase that it modifies:

> Because he was one of the greatest thinkers of Renaissance Italy, John idolized Leonardo da Vinci.

In this sentence, the modifier is "because he was one of the greatest thinkers of Renaissance Italy," and the noun it is intended to modify is "Leonardo da Vinci." However, due to the placement of the modifier next to the subject, John, it seems as if the sentence is stating that John was a Renaissance genius, not Da Vinci.

> John idolized Leonard da Vinci because he was one of the greatest thinkers of Renaissance Italy.

The modifier is now adjacent to the appropriate noun, clarifying which of the two men in this sentence is the greatest thinker.

Dangling modifiers modify a word or phrase that is not readily apparent in the sentence. That is, they "dangle" because they are not clearly attached to anything:

> After getting accepted to college, Amir's parents were proud.

The modifier here, "after getting accepted to college," should modify who got accepted. The noun immediately following the modifier is "Amir's parents"—but they are probably not the ones who are going to college.

> After getting accepted to college, Amir made his parents proud.

The subject of the sentence has been changed to Amir himself, and now the subject and its modifier are appropriately matched.

Inconsistent Verb Tense

Verb tense reflects when an action occurred or a state existed. For example, the tense known as *simple present* expresses something that is happening right now or that happens regularly:

> She *works* in a hospital.

Present continuous tense expresses something in progress. It is formed by to be + verb + -ing.

> Sorry, I can't go out right now. I *am doing* my homework.

Past tense is used to describe events that previously occurred. However, in conversational English, speakers often use present tense or a mix of past and present tense when relating past events because it gives the narrative a sense of immediacy. In formal written English, though, consistency in verb tense is necessary to avoid reader confusion.

> I traveled to Europe last summer. As soon as I stepped off the plane, I feel like I'm in a movie! I'm surrounded by quaint cafes and impressive architecture.

The passage above abruptly switches from past tense—*traveled, stepped*—to present tense—*feel, am surrounded*.

> I *traveled* to Europe last summer. As soon as I *stepped* off the plane, I *felt* like I was in a movie! I *was surrounded* by quaint cafes and impressive architecture.

All verbs are in past tense, so this passage now has consistent verb tense.

Split Infinitives

The *infinitive form* of a verb consists of "to + base verb"—e.g., to walk, to sleep, to approve. A *split infinitive* occurs when another word, usually an adverb, is placed between *to* and the verb:

> I decided *to simply walk* to work to get more exercise every day.

The infinitive *to walk* is split by the adverb *simply*.

> It was a mistake *to hastily approve* the project before conducting further preliminary research.

The infinitive *to approve* is split by *hastily*.

Although some grammarians still advise against split infinitives, this syntactic structure is common in both spoken and written English and is widely accepted in standard usage.

Subject-Verb Agreement

In English, verbs must agree with the subject. The form of a verb may change depending on whether the subject is singular or plural, or whether it is first, second, or third person. For example, the verb *to be* has various forms:

> I am a student.

> You are a student.

> She is a student.

> We are students.

> They are students.

Errors occur when a verb does not agree with its subject. Sometimes, the error is readily apparent:

> We is hungry.

Is is not the appropriate form of *to be* when used with the third person plural *we*.

> We are hungry.

This sentence now has correct subject-verb agreement.

However, some cases are trickier, particularly when the subject consists of a lengthy noun phrase with many modifiers:

> Students who are hoping to accompany the anthropology department on its annual summer trip to Ecuador needs to sign up by March 31st.

The verb in this sentence is *needs*. However, its subject is not the noun adjacent to it—Ecuador. The subject is the noun at the beginning of the sentence—students. Because *students* is plural, *needs* is the incorrect verb form.

> *Students* who are hoping to accompany the anthropology department on its annual summer trip to Ecuador *need* to sign up by March 31st.

This sentence now uses correct agreement between *students* and *need*.

Another case to be aware of is a *collective noun*. A collective noun refers to a group of many things or people but can be singular in itself—e.g., family, committee, army, pair team, council, jury. Whether or not a collective noun uses a singular or plural verb depends on how the noun is being used. If the noun refers to the group performing a collective action as one unit, it should use a singular verb conjugation:

> The family is moving to a new neighborhood.

The whole family is moving together in unison, so the singular verb form *is* is appropriate here.

> The committee has made its decision.

The verb *has* and the possessive pronoun *its* both reflect the word *committee* as a singular noun in the sentence above; however, when a collective noun refers to the group as individuals, it can take a plural verb:

> The newlywed pair spend every moment together.

This sentence emphasizes the love between two people in a pair, so it can use the plural verb *spend*.

> The council are all newly elected members.

The sentence refers to the council in terms of its individual members and uses the plural verb *are*.

Overall though, American English is more likely to pair a collective noun with a singular verb, while British English is more likely to pair a collective noun with a plural verb.

Grammar, Usage, Syntax, and Mechanics Choices

Colons and Semicolons

In a sentence, *colons* are used before a list, a summary or elaboration, or an explanation related to the preceding information in the sentence:

> There are two ways to reserve tickets for the performance: by phone or in person.

> One thing is clear: students are spending more on tuition than ever before.

As these examples show, a colon must be preceded by an independent clause. However, the information after the colon may be in the form of an independent clause or in the form of a list.

Semicolons can be used in two different ways—to join ideas or to separate them. In some cases, semicolons can be used to connect what would otherwise be stand-alone sentences. Each part of the sentence joined by a semicolon must be an independent clause. The use of a semicolon indicates that these two independent clauses are closely related to each other:

> The rising cost of childcare is one major stressor for parents; healthcare expenses are another source of anxiety.

> Classes have been canceled due to the snowstorm; check the school website for updates.

Semicolons can also be used to divide elements of a sentence in a more distinct way than simply using a comma. This usage is particularly useful when the items in a list are especially long and complex and contain other internal punctuation.

> Retirees have many modes of income: some survive solely off their retirement checks; others supplement their income through part time jobs, like working in a supermarket or substitute teaching; and others are financially dependent on the support of family members, friends, and spouses.

Its and It's

These pronouns are some of the most confused in the English language as most possessives contain the suffix –'s. However, for *it*, it is the opposite. *Its* is a possessive pronoun:

> The government is reassessing *its* spending plan.

It's is a contraction of the words *it is*:

> *It's* snowing outside.

Saw and Seen

Saw and *seen* are both conjugations of the verb *to see*, but they express different verb tenses. *Saw* is used in the simple past tense. *Seen* is the past participle form of *to see* and can be used in all perfect tenses.

> I seen her yesterday.

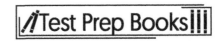

This sentence is incorrect. Because it expresses a completed event from a specified point in time in the past, it should use simple past tense:

I *saw* her yesterday.

This sentence uses the correct verb tense. Here's how the past participle is used correctly:

I *have seen* her before.

The meaning in this sentence is slightly changed to indicate an event from an unspecific time in the past. In this case, present perfect is the appropriate verb tense to indicate an unspecified past experience. Present perfect conjugation is created by combining *to have* + past participle.

Then and Than

Then is generally used as an adverb indicating something that happened next in a sequence or as the result of a conditional situation:

We parked the car and *then* walked to the restaurant.

If enough people register for the event, *then* we can begin planning.

Than is a conjunction indicating comparison:

This watch is more expensive *than* that one.

The bus departed later *than* I expected.

They're, Their, and There

They're is a contraction of the words *they are*:

They're moving to Ohio next week.

Their is a possessive pronoun:

The baseball players are training for *their* upcoming season.

There can function as multiple parts of speech, but it is most commonly used as an adverb indicating a location:

Let's go to the concert! Some great bands are playing *there*.

Insure and Ensure

These terms are both verbs. *Insure* means to guarantee something against loss, harm, or damage, usually through an insurance policy that offers monetary compensation:

The robbers made off with her prized diamond necklace, but luckily it was *insured* for one million dollars.

Ensure means to make sure, to confirm, or to be certain:

Ensure that you have your passport before entering the security checkpoint.

Accept and Except

Accept is a verb meaning to take or agree to something:

> I would like to *accept* your offer of employment.

Except is a preposition that indicates exclusion:

> I've been to every state in America *except* Hawaii.

Affect and Effect

Affect is a verb meaning to influence or to have an impact on something:

> The amount of rainfall during the growing season *affects* the flavor of wine produced from these grapes.

Effect can be used as either a noun or a verb. As a noun, *effect* is synonymous with a result:

> If we implement the changes, what will the *effect* be on our profits?

As a verb, *effect* means to bring about or to make happen:

> In just a few short months, the healthy committee has *effected* real change in school nutrition.

Components of Sentences

Clauses

Clauses contain a subject and a verb. An *independent clause* can function as a complete sentence on its own, but it might also be one component of a longer sentence. *Dependent clauses* cannot stand alone as complete sentences. They rely on independent clauses to complete their meaning. Dependent clauses usually begin with a subordinating conjunction. Independent and dependent clauses are sometimes also referred to as *main clauses* and *subordinate clauses*, respectively. The following structure highlights the differences:

> Apiculturists raise honeybees because they love insects.

Apiculturists raise honeybees is an independent or main clause. The subject is *apiculturists*, and the verb is *raise*. It expresses a complete thought and could be a standalone sentence.

Because they love insects is a dependent or subordinate clause. If it were not attached to the independent clause, it would be a sentence fragment. While it contains a subject and verb—*they love*—this clause is dependent because it begins with the subordinate conjunction *because*. Thus, it does not express a complete thought on its own.

Another type of clause is a *relative clause*, and it is sometimes referred to as an *adjective clause* because it gives further description about the noun. A relative clause begins with a *relative pronoun*: *that, which, who, whom, whichever, whomever,* or *whoever*. It may also begin with a *relative adverb*: *where, why,* or *when*. Here's an example of a relative clause, functioning as an adjective:

> The strawberries that I bought yesterday are already beginning to spoil.

Here, the relative clause is *that I bought yesterday*; the relative pronoun is *that*. The subject is *I*, and the verb is *bought*. The clause modifies the subject *strawberries* by answering the question, "Which strawberries?" Here's an example of a relative clause with an adverb:

The tutoring center is a place where students can get help with homework.

The relative clause is *where students can get help with homework*, and it gives more information about a place by describing what kind of place it is. It begins with the relative adverb *where* and contains the noun *students* along with its verb phrase *can get*.

Relative clauses may be further divided into two types: essential or nonessential. *Essential clauses* contain identifying information without which the sentence would lose significant meaning or not make sense. These are also sometimes referred to as *restrictive clauses*. The sentence above contains an example of an essential relative clause. Here is what happens when the clause is removed:

The tutoring center is a place where students can get help with homework.

The tutoring center is a place.

Without the relative clause, the sentence loses the majority of its meaning; thus, the clause is essential or restrictive.

Nonessential clauses—also referred to as *non-restrictive clauses*—offer additional information about a noun in the sentence, but they do not significantly control the overall meaning of the sentence. The following example indicates a nonessential clause:

New York City, which is located in the northeastern part of the country, is the most populated city in America.

New York City is the most populated city in America.

Even without the relative clause, the sentence is still understandable and continues to communicate its central message about New York City. Thus, it is a nonessential clause.

Punctuation differs between essential and nonessential relative clauses, too. Nonessential clauses are set apart from the sentence using commas whereas essential clauses are not separated with commas. Also, the relative pronoun *that* is generally used for essential clauses, while *which* is used for nonessential clauses. The following examples clarify this distinction:

Romeo and Juliet is my favorite play *that Shakespeare wrote*.

The relative clause *that Shakespeare wrote* contains essential, controlling information about the noun *play*, limiting it to those plays by Shakespeare. Without it, it would seem that *Romeo and Juliet* is the speaker's favorite play out of every play ever written, not simply from Shakespeare's repertoire.

Romeo and Juliet, which Shakespeare wrote, is my favorite play.

Here, the nonessential relative clause—"which Shakespeare wrote"—modifies *Romeo and Juliet*. It doesn't provide controlling information about the play, but simply offers further background details. Thus, commas are needed.

Phrases

Phrases are groups of words that do not contain the subject-verb combination required for clauses. Phrases are classified by the part of speech that begins or controls the phrase.

A *noun phrase* consists of a noun and all its modifiers—adjectives, adverbs, and determiners. Noun phrases can serve many functions in a sentence, acting as subjects, objects, and object complements:

> *The shallow yellow bowl* sits on the top shelf.

> Nina just bought *some incredibly fresh organic produce*.

Prepositional phrases are made up of a preposition and its object. The object of a preposition might be a noun, noun phrase, pronoun, or gerund. Prepositional phrases may function as either an adjective or an adverb:

> Jack picked up the book *in front of him*.

The prepositional phrase *in front of him* acts as an adjective indicating which book Jack picked up.

> The dog ran into the back yard.

The phrase *into the backyard* describes where the dog ran, so it acts as an adverb.

Verb phrases include all of the words in a verb group, even if they are not directly adjacent to each other:

> I *should have woken up* earlier this morning.

> The company **is** now *offering* membership discounts for new enrollers.

This sentence's verb phrase is *is offering*. Even though they are separated by the word *now*, they function together as a single verb phrase.

Structures of Sentences

All sentences contain the same basic elements: a subject and a verb. The *subject* is who or what the sentence is about; the *verb* describes the subject's action or condition. However, these elements, subjects and verbs, can be combined in different ways. The following graphic describes the different types of sentence structures.

Sentence Structure	Independent Clauses	Dependent Clauses
Simple	1	0
Compound	2 or more	0
Complex	1	1 or more
Compound-Complex	2 or more	1 or more

A *simple sentence* expresses a complete thought and consists of one subject and verb combination:

> The children ate pizza.

The subject is *children*. The verb is *ate*.

Either the subject or the verb may be *compound*—that is, it could have more than one element:

> *The children and their parents* ate pizza.

> The children *ate pizza and watched a movie.*

All of these are still simple sentences. Despite having either compound subjects or compound verbs, each sentence still has only one subject and verb combination.

Compound sentences combine two or more simple sentences to form one sentence that has multiple subject-verb combinations:

> *The children ate pizza,* and *their parents watched a movie.*

This structure is comprised of two independent clauses: (1) *the children ate pizza* and (2) *their parents watched a movie.* Compound sentences join different subject-verb combinations using a comma and a coordinating conjunction.

> I called my mom**,** *but* she didn't answer the phone.

> The weather was stormy, *so* we canceled our trip to the beach.

A *complex sentence* consists of an independent clause and one or more dependent clauses. Dependent clauses join a sentence using *subordinating conjunctions*. Some examples of subordinating conjunctions are *although, unless, as soon as, since, while, when, because, if,* and *before*.

> I missed class yesterday *because* my mother was ill.

> *Before* traveling to a new country, you need to exchange your money to the local currency.

The order of clauses determines their punctuation. If the dependent clause comes first, it should be separated from the independent clause with a comma. However, if the complex sentence consists of an independent clause followed by a dependent clause, then a comma is not always necessary.

A *compound-complex sentence* can be created by joining two or more independent clauses with at least one dependent clause:

> After the earthquake struck, thousands of homes were destroyed, and many families were left without a place to live.

The first independent clause in the compound structure includes a subordinating clause—*after the earthquake struck*. Thus, the structure is both complex and compound.

Understanding the Use of Affixes, Context, and Syntax

Affixes

Individual words are constructed from building blocks of meaning. An *affix* is an element that is added to a root or stem word that can change the word's meaning.

For example, the stem word *fix* is a verb meaning *to repair*. When the ending *–able* is added, it becomes the adjective *fixable*, meaning "capable of being repaired." Adding *un–* to the beginning changes the word to *unfixable*, meaning "incapable of being repaired." In this way, affixes attach to the word stem to create a new word and a new meaning. Knowledge of affixes can assist in deciphering the meaning of unfamiliar words.

Affixes are also related to inflection. *Inflection* is the modification of a base word to express a different grammatical or syntactical function. For example, countable nouns such as *car* and *airport* become plural with the addition of *–s* at the end: *cars* and *airports*.

Verb tense is also expressed through inflection. *Regular verbs*—those that follow a standard inflection pattern—can be changed to past tense using the affixes *–ed*, *–d*, or *–ied*, as in *cooked* and *studied*. Verbs can also be modified for continuous tenses by using *–ing*, as in *working* or *exploring*. Thus, affixes are used not only to express meaning but also to reflect a word's grammatical purpose.

A *prefix* is an affix attached to the beginning of a word. The meanings of English prefixes mainly come from Greek and Latin origins. The chart below contains a few of the most commonly used English prefixes.

Prefix	Meaning	Example
a-	not	amoral, asymptomatic
anti-	against	antidote, antifreeze
auto-	self	automobile, automatic
circum-	around	circumference, circumspect
co-, com-, con-	together	coworker, companion
contra-	against	contradict, contrary
de-	negation or reversal	deflate, deodorant
extra-	outside, beyond	extraterrestrial, extracurricular
in-, im-, il-, ir-	not	impossible, irregular

Prefix	Meaning	Example
inter-	between	international, intervene
intra-	within	intramural, intranet
mis-	wrongly	mistake, misunderstand
mono-	one	monolith, monopoly
non-	not	nonpartisan, nonsense
pre-	before	preview, prediction
re-	again	review, renew
semi-	half	semicircle, semicolon
sub-	under	subway, submarine
super-	above	superhuman, superintendent
trans-	across, beyond, through	trans-Siberian, transform
un-	not	unwelcome, unfriendly

While the addition of a prefix alters the meaning of the base word, the addition of a *suffix* may also affect a word's part of speech. For example, adding a suffix can change the noun *material* into the verb *materialize* and back to a noun again in *materialization*.

Suffix	Part of Speech	Meaning	Example
-able, -ible	adjective	having the ability to	honorable, flexible
-acy, -cy	noun	state or quality	intimacy, dependency
-al, -ical	adjective	having the quality of	historical, tribal
-en	verb	to cause to become	strengthen, embolden
-er, -ier	adjective	comparative	happier, longer
-est, -iest	adjective	superlative	sunniest, hottest
-ess	noun	female	waitress, actress
-ful	adjective	full of, characterized by	beautiful, thankful
-fy, -ify	verb	to cause, to come to be	liquefy, intensify
-ism	noun	doctrine, belief, action	Communism, Buddhism
-ive, -ative, -itive	adjective	having the quality of	creative, innovative
-ize	verb	to convert into, to subject to	Americanize, dramatize
-less	adjective	without, missing	emotionless, hopeless
-ly	adverb	in the manner of	quickly, energetically
-ness	noun	quality or state	goodness, darkness
-ous, -ious, -eous	adjective	having the quality of	spontaneous, pious
-ship	noun	status or condition	partnership, ownership
-tion	noun	action or state	renovation, promotion
-y	adjective	characterized by	smoky, dreamy

Through knowledge of prefixes and suffixes, a student's vocabulary can be instantly expanded with an understanding of *etymology*—the origin of words. This, in turn, can be used to add sentence structure variety to academic writing.

Context Clues

Familiarity with common prefixes, suffixes, and root words assists tremendously in unraveling the meaning of an unfamiliar word and making an educated guess as to its meaning. However, some words do not contain many easily-identifiable clues that point to their meaning. In this case, rather than looking at the elements within the word, it is useful to consider elements around the word—i.e., its context. *Context* refers to the other words and information within the sentence or surrounding sentences that indicate the unknown word's probable meaning. The following sentences provide context for the potentially-unfamiliar word *quixotic*:

> Rebecca had never been one to settle into a predictable, ordinary life. Her quixotic personality led her to leave behind a job with a prestigious law firm in Manhattan and move halfway around the world to pursue her dream of becoming a sushi chef in Tokyo.

A reader unfamiliar with the word *quixotic* doesn't have many clues to use in terms of affixes or root meaning. The suffix *-ic* indicates that the word is an adjective, but that is it. In this case, then, a reader would need to look at surrounding information to obtain some clues about the word. Other adjectives in the passage include *predictable* and *ordinary*, things that Rebecca was definitely not, as indicated by "Rebecca had never been one to settle." Thus, a first clue might be that *quixotic* means the opposite of predictable.

The second sentence doesn't offer any other modifier of *personality* other than *quixotic*, but it does include a story that reveals further information about her personality. She had a stable, respectable job, but she decided to give it up to follow her dream. Combining these two ideas together, then—unpredictable and dream-seeking—gives the reader a general idea of what *quixotic* probably means. In fact, the root of the word is the character Don Quixote, a romantic dreamer who goes on an impulsive adventure.

While context clues are useful for making an approximate definition for newly-encountered words, these types of clues also come in handy when encountering common words that have multiple meanings. The word *reservation* is used differently in each the following sentences:

A. That restaurant is booked solid for the next month; it's impossible to make a reservation unless you know somebody.

B. The hospital plans to open a branch office inside the reservation to better serve Native American patients who cannot easily travel to the main hospital fifty miles away.

C. Janet Clark is a dependable, knowledgeable worker, and I recommend her for the position of team leader without reservation.

All three sentences use the word to express different meanings. In fact, most words in English have more than one meaning—sometimes meanings that are completely different from one another. Thus, context can provide clues as to which meaning is appropriate in a given situation. A quick search in the dictionary reveals several possible meanings for *reservation*:

1. An exception or qualification
2. A tract of public land set aside, such as for the use of American Indian tribes
3. An arrangement for accommodations, such as in a hotel, on a plane, or at a restaurant

Sentence A mentions a restaurant, making the third definition the correct one in this case. In sentence B, some context clues include Native Americans, as well as the implication that a reservation is a place—"inside the reservation," both of which indicate that the second definition should be used here. Finally, sentence C uses *without reservation* to mean "completely" or "without exception," so the first definition can be applied here.

Using context clues in this way can be especially useful for words that have multiple, widely varying meanings. If a word has more than one definition and two of those definitions are the opposite of each other, it is known as an *auto-antonym*—a word that can also be its own antonym. In the case of auto-antonyms, context clues are crucial to determine which definition to employ in a given sentence. For example, the word *sanction* can either mean "to approve or allow" or "a penalty." Approving and penalizing have opposite meanings, so *sanction* is an example of an auto-antonym. The following sentences reflect the distinction in meaning:

A. In response to North Korea's latest nuclear weapons test, world leaders have called for harsher sanctions to punish the country for its actions.

B. The general has sanctioned a withdrawal of troops from the area.

A context clue can be found in sentence A, which mentions "to punish." A punishment is similar to a penalty, so sentence A is using the word *sanction* according to this definition.

Other examples of auto-antonyms include *oversight*—"to supervise something" or "a missed detail"), *resign*—"to quit" or "to sign again, as a contract," and *screen*—"to show" or "to conceal." For these types of words, recognizing context clues is an important way to avoid misinterpreting the sentence's meaning.

Syntax

Syntax refers to the arrangement of words, phrases, and clauses to form a sentence. Knowledge of syntax can also give insight into a word's meaning. The section above considered several examples using the word *reservation* and applied context clues to determine the word's appropriate meaning in each sentence. Here is an example of how the placement of a word can impact its meaning and grammatical function:

A. The development team has reserved the conference room for today.

B. Her quiet and reserved nature is sometimes misinterpreted as unfriendliness when people first meet her.

In addition to using *reserved* to mean different things, each sentence also uses the word to serve a different grammatical function. In sentence A, *reserved* is part of the verb phrase *has reserved*, indicating the meaning "to set aside for a particular use." In sentence B, *reserved* acts as a modifier within the noun phrase "her quiet and reserved nature." Because the word is being used as an adjective to describe a personality characteristic, it calls up a different definition of the word—"restrained or lacking familiarity with others." As this example shows, the function of a word within the overall sentence structure can allude to its meaning. It is also useful to refer to the earlier chart about suffixes and parts of speech as another clue into what grammatical function a word is serving in a sentence.

Analyzing Nuances of Word Meaning and Figures of Speech

By now, it should be apparent that language is not as simple as one word directly correlated to one meaning. Rather, one word can express a vast array of diverse meanings, and similar meanings can be expressed through different words. However, there are very few words that express exactly the same meaning. For this reason, it is important to be able to pick up on the nuances of word meaning.

Many words contain two levels of meaning: connotation and denotation as discussed previously in the informational texts and rhetoric section. A word's *denotation* is its most literal meaning—the definition that can readily be found in the dictionary. A word's *connotation* includes all of its emotional and cultural associations.

In literary writing, authors rely heavily on connotative meaning to create mood and characterization. The following are two descriptions of a rainstorm:

A. The rain slammed against the windowpane and the wind howled through the fireplace. A pair of hulking oaks next to the house cast eerie shadows as their branches trembled in the wind.

B. The rain pattered against the windowpane and the wind whistled through the fireplace. A pair of stately oaks next to the house cast curious shadows as their branches swayed in the wind.

Description A paints a creepy picture for readers with strongly emotional words like *slammed*, connoting force and violence. *Howled* connotes pain or wildness, and *eerie* and *trembled* connote fear. Overall, the connotative language in this description serves to inspire fear and anxiety.

However, as can be seen in description B, swapping out a few key words for those with different connotations completely changes the feeling of the passage. *Slammed* is replaced with the more cheerful *pattered*, and *hulking* has been swapped out for *stately*. Both words imply something large, but *hulking* is more intimidating whereas *stately* is more respectable. *Curious* and *swayed* seem more playful than the language used in the earlier description. Although both descriptions represent roughly the same situation, the nuances of the emotional language used throughout the passages create a very different sense for readers.

Selective choice of connotative language can also be extremely impactful in other forms of writing, such as editorials or persuasive texts. Through connotative language, writers reveal their biases and opinions while trying to inspire feelings and actions in readers:

A. Parents won't stop complaining about standardized tests.
B. Parents continue to raise concerns about standardized tests.

Readers should be able to identify the nuance in meaning between these two sentences. The first one carries a more negative feeling, implying that parents are being bothersome or whiny. Readers of the second sentence, though, might come away with the feeling that parents are concerned and involved in their children's education. Again, the aggregate of even subtle cues can combine to give a specific emotional impression to readers, so from an early age, students should be aware of how language can be used to influence readers' opinions.

Another form of non-literal expression can be found in *figures of speech*. As with connotative language, figures of speech tend to be shared within a cultural group and may be difficult to pick up on for learners outside of that group. In some cases, a figure of speech may be based on the literal denotation of the

58

words it contains, but in other cases, a figure of speech is far removed from its literal meaning. A case in point is *irony*, where what is said is the exact opposite of what is meant:

> The new tax plan is poorly planned, based on faulty economic data, and unable to address the financial struggles of middle class families. Yet legislators remain committed to passing this brilliant proposal.

When the writer refers to the proposal as brilliant, the opposite is implied—the plan is "faulty" and "poorly planned." By using irony, the writer means that the proposal is anything but brilliant by using the word in a non-literal sense.

Another figure of speech is *hyperbole*—extreme exaggeration or overstatement. Statements like, "I love you to the moon and back" or "Let's be friends for a million years" utilize hyperbole to convey a greater depth of emotion, without literally committing oneself to space travel or a life of immortality.

Figures of speech may sometimes use one word in place of another. *Synecdoche*, for example, uses a part of something to refer to its whole. The expression "Don't hurt a hair on her head!" implies protecting more than just an individual hair, but rather her entire body. "The art teacher is training a class of Picassos" uses Picasso, one individual notable artist, to stand in for the entire category of talented artists. Another figure of speech using word replacement is *metonymy*, where a word is replaced with something closely associated to it. For example, news reports may use the word "Washington" to refer to the American government or "the crown" to refer to the British monarch.

Practice Questions

1. Which sentence contains an error in punctuation or capitalization?
    a. "The show is on," Jackson said.
    b. The Grand Canyon is a national park.
    c. Lets celebrate tomorrow.
    d. Oliver, a social worker, got a new job this month.

2. Which of the following sentences contains an error in usage?
    a. Their words was followed by a signing document.
    b. No one came to the theater that evening.
    c. Several cats were living in the abandoned house down the road.
    d. It rained that morning; they had to cancel the kayaking trip.

3. What type of grammatical error does the following sentence contain?

    It was true, Lyla ate the last cupcake.

    a. Subject-verb agreement error
    b. Punctuation error
    c. Shift in verb tense
    d. Split infinitive

4. Which sentence that contains an error in punctuation or capitalization?
    a. Afterwards, we got ice cream down the road.
    b. The word "slacken" means to decrease.
    c. They started building the Hoover dam in 1931.
    d. Matthew got married to his best friend, Maria.

5. Which sentence contains an error in usage?
    a. After her swim, Jeanine saw a blue kid's shovel.
    b. Pistachios are my favorite kind of nut, although they're expensive.
    c. One apple is better than two lemons.
    d. We found three five-dollar bills on the way home.

Answer Explanations

1. C: The correct answer choice is "Lets celebrate tomorrow." "Lets" is supposed to be short for "let us," and therefore needs an apostrophe between the "t" and the "s": "Let's."

2. A: This error is marked by a subject/verb agreement. "Words" is plural, so the verb must be plural as well. The correct usage would be: "Their words were followed by a signing document."

3. B: There is a punctuation error. The comma creates a comma splice where a period or a semicolon should be, since we have two independent clauses on either side of the comma.

4. C: Choice *C* is the problematic answer; the whole phrase "Hoover Dam" should be capitalized, not just "Hoover."

5. A: Choice *A* has the error in usage because we have a dangling modifier with the phrase "blue kid's shovel." The sentence indicates the kid is blue. We want the sentence to say that the shovel is blue. Therefore, it should be: "After her swim, Jeanine saw a kid's blue shovel."

Math

Number Sense and Operations

Base-10 Numerals, Number Names, and Expanded Form

Numbers used in everyday life are constituted in a base-10 system. Each digit in a number, depending on its location, represents some multiple of 10, or quotient of 10 when dealing with decimals. Each digit to the left of the decimal point represents a higher multiple of 10. Each digit to the right of the decimal point represents a quotient of a higher multiple of 10 for the divisor. For example, consider the number 7,631.42. The digit one represents simply the number one. The digit 3 represents 3×10.

The digit 6 represents $6 \times 10 \times 10$ (or 6×100).

The digit 7 represents $7 \times 10 \times 10 \times 10$ (or 7×1000).

The digit 4 represents $4 \div 10$.

The digit 2 represents $(2 \div 10) \div 10$, or $2 \div (10 \times 10)$ or $2 \div 100$.

A number is written in expanded form by expressing it as the sum of the value of each of its digits. The expanded form in the example above, which is written with the highest value first down to the lowest value, is expressed as:

$$7,000 + 600 + 30 + 1 + .4 + .02$$

When verbally expressing a number, the integer part of the number (the numbers to the left of the decimal point) resembles the expanded form without the addition between values. In the above example, the numbers read "seven thousand six hundred thirty-one." When verbally expressing the decimal portion of a number, the number is read as a whole number, followed by the place value of the furthest digit (non-zero) to the right. In the above example, 0.42 is read "forty-two hundredths." Reading the number 7,631.42 in its entirety is expressed as "seven thousand six hundred thirty-one and forty-two hundredths." The word *and* is used between the integer and decimal parts of the number.

Composing and Decomposing Multi-Digit Numbers

Composing and decomposing numbers aids in conceptualizing what each digit of a multi-digit number represents. The standard, or typical, form in which numbers are written consists of a series of digits representing a given value based on their place value. Consider the number 592.7. This number is composed of 5 hundreds, 9 tens, 2 ones, and 7 tenths.

Composing a number requires adding the given numbers for each place value and writing the numbers in standard form. For example, composing 4 thousands, 5 hundreds, 2 tens, and 8 ones consists of adding as follows: $4,000 + 500 + 20 + 8$, to produce 4,528 (standard form).

Decomposing a number requires taking a number written in standard form and breaking it apart into the sum of each place value. For example, the number 83.17 is decomposed by breaking it into the sum of 4 values (for each of the 4 digits): 8 tens, 3 ones, 1 tenth, and 7 hundredths. The decomposed or "expanded" form of 83.17 is $80 + 3 + .1 + .07$.

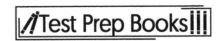

Place Value of a Given Digit

The number system that is used consists of only ten different digits or characters. However, this system is used to represent an infinite number of values. The place value system makes this infinite number of values possible. The position in which a digit is written corresponds to a given value. Starting from the decimal point (which is implied, if not physically present), each subsequent place value to the left represents a value greater than the one before it. Conversely, starting from the decimal point, each subsequent place value to the right represents a value less than the one before it.

The names for the place values to the left of the decimal point are as follows:

...	Billions	Hundred-Millions	Ten-Millions	Millions	Hundred-Thousands	Ten-Thousands	Thousands	Hundreds	Tens	Ones

*Note that this table can be extended infinitely further to the left.

The names for the place values to the right of the decimal point are as follows:

Decimal Point (.)	Tenths	Hundredths	Thousandths	Ten-Thousandths	...

*Note that this table can be extended infinitely further to the right.

When given a multi-digit number, the value of each digit depends on its place value. Consider the number 682,174.953. Referring to the chart above, it can be determined that the digit 8 is in the ten-thousands place. It is in the fifth place to the left of the decimal point. Its value is 8 ten-thousands or 80,000. The digit 5 is two places to the right of the decimal point. Therefore, the digit 5 is in the hundredths place. Its value is 5 hundredths or $\frac{5}{100}$ (equivalent to .05).

Base-10 System

Value of Digits

In accordance with the base-10 system, the value of a digit increases by a factor of ten each place it moves to the left. For example, consider the number 7. Moving the digit one place to the left (70), increases its value by a factor of 10 ($7 \times 10 = 70$).

Moving the digit two places to the left (700) increases its value by a factor of 10 twice

$$(7 \times 10 \times 10 = 700)$$

Moving the digit three places to the left (7,000) increases its value by a factor of 10 three times ($7 \times 10 \times 10 \times 10 = 7,000$), and so on.

Conversely, the value of a digit decreases by a factor of ten each place it moves to the right. (Note that multiplying by $\frac{1}{10}$ is equivalent to dividing by 10). For example, consider the number 40. Moving the digit one place to the right (4) decreases its value by a factor of 10 ($40 \div 10 = 4$).

Moving the digit two places to the right (0.4), decreases its value by a factor of 10 twice ($40 \div 10 \div 10 = 0.4$) or ($40 \times \frac{1}{10} \times \frac{1}{10} = 0.4$).

Moving the digit three places to the right (0.04) decreases its value by a factor of 10 three times:

$$(40 \div 10 \div 10 \div 10 = 0.04)$$

or

$$(40 \times \frac{1}{10} \times \frac{1}{10} \times \frac{1}{10} = 0.04) \text{ and so on.}$$

Exponents to Denote Powers of 10

The value of a given digit of a number in the base-10 system can be expressed utilizing powers of 10. A power of 10 refers to 10 raised to a given exponent such as 10^0, 10^1, 10^2, 10^3, etc.

For the number 10^3, 10 is the base and 3 is the exponent. A base raised by an exponent represents how many times the base is multiplied by itself. Therefore,

$$10^1 = 10$$

$$10^2 = 10 \times 10 = 100$$

$$10^3 = 10 \times 10 \times 10 = 1,000$$

$$10^4 = 10 \times 10 \times 10 \times 10 = 10,000, \text{ etc.}$$

Any base with a zero exponent equals one.

Powers of 10 are utilized to decompose a multi-digit number without writing all the zeroes. Consider the number 872,349. This number is decomposed to $800,000 + 70,000 + 2,000 + 300 + 40 + 9$. When utilizing powers of 10, the number 872,349 is decomposed to:

$$(8 \times 10^5) + (7 \times 10^4) + (2 \times 10^3) + (3 \times 10^2) + (4 \times 10^1) + (9 \times 10^0)$$

The power of 10 by which the digit is multiplied corresponds to the number of zeroes following the digit when expressing its value in standard form. For example, 7×10^4 is equivalent to 70,000 or 7 followed by four zeros.

Comparing, Classifying, and Ordering Rational Numbers

A **rational number** is any number that can be written as a fraction or ratio. Within the set of rational numbers, several subsets exist that are referenced throughout the mathematics topics. Counting numbers are the first numbers learned as a child. Counting numbers consist of 1,2,3,4, and so on. Whole numbers include all counting numbers and zero (0,1,2,3,4,...). Integers include counting numbers, their opposites, and zero (..., -3, -2, -1, 0, 1, 2, 3, ...). Rational numbers are inclusive of integers, fractions, and decimals that terminate, or end (1.7, 0.04213) or repeat (0.136$\bar{5}$).

When comparing or ordering numbers, the numbers should be written in the same format (decimal or fraction), if possible. For example, $\sqrt{49}$, 7.3, and $\frac{15}{2}$ are easier to order if each one is converted to a decimal, such as 7, 7.3, and 7.5 (converting fractions and decimals is covered in the following section). A number line is used to order and compare the numbers. Any number that is to the right of another number is greater than that number. Conversely, a number positioned to the left of a given number is less than that number.

Structure of the Number System

The mathematical number system is made up of two general types of numbers: real and complex. *Real numbers* are those that are used in normal settings, while *complex numbers* are those composed of both a real number and an imaginary one. Imaginary numbers are the result of taking the square root of -1, and $\sqrt{-1} = i$.

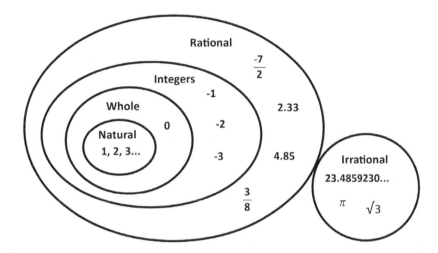

The real number system is often explained using a Venn diagram similar to the one below. After a number has been labeled as a real number, further classification occurs when considering the other groups in this diagram. If a number is a never-ending, non-repeating decimal, it falls in the irrational category. Otherwise, it is rational. More information on these types of numbers is provided in the previous section. Furthermore, if a number does not have a fractional part, it is classified as an integer, such as -2, 75, or 0. Whole numbers are an even smaller group that only includes positive integers and 0. The last group of natural numbers is made up of only positive integers, such as 2, 56, or 12.

Real numbers can be compared and ordered using the number line. If a number falls to the left on the real number line, it is less than a number on the right. For example, $-2 < 5$ because -2 falls to the left of 0, and 5 falls to the right. Numbers to the left of zero are negative while those to the right are positive.

Complex numbers are made up of the sum of a real number and an imaginary number. Some examples of complex numbers include:

$$6 + 2i$$

$$5 - 7i$$

$$-3 + 12i$$

Adding and subtracting complex numbers is similar to collecting like terms. The real numbers are added together, and the imaginary numbers are added together. For example, if the problem asks to simplify the expression:

$$6 + 2i - 3 + 7i$$

the 6 and -3 are combined to make 3, and the $2i$ and $7i$ combine to make $9i$. Multiplying and dividing complex numbers is similar to working with exponents.

One rule to remember when multiplying is that $i * i = -1$. For example, if a problem asks to simplify the expression:

$$4i(3 + 7i)$$

the $4i$ should be distributed throughout the 3 and the $7i$. This leaves the final expression $12i - 28$. The 28 is negative because $i * i$ results in a negative number. The last type of operation to consider with complex numbers is the conjugate. The *conjugate* of a complex number is a technique used to change the complex number into a real number. For example, the conjugate of $4 - 3i$ is $4 + 3i$. Multiplying $(4 - 3i)(4 + 3i)$ results in:

$$16 + 12i - 12i + 9$$

which has a final answer of $16 + 9 = 25$.

The order of operations—PEMDAS—simplifies longer expressions with real or imaginary numbers. Each operation is listed in the order of how they should be completed in a problem containing more than one operation. Parenthesis can also mean grouping symbols, such as brackets and absolute value. Then, exponents are calculated. Multiplication and division should be completed from left to right, and addition and subtraction should be completed from left to right.

Simplification of another type of expression occurs when radicals are involved. As explained previously, root is another word for radical. For example, the following expression is a radical that can be simplified: $\sqrt{24x^2}$. First, the number must be factored out to the highest perfect square. Any perfect square can be taken out of a radical. Twenty-four can be factored into 4 and 6, and 4 can be taken out of the radical. $\sqrt{4} = 2$ can be taken out, and 6 stays underneath. If $x > 0$, x can be taken out of the radical because it is a perfect square. The simplified radical is $2x\sqrt{6}$. An approximation can be found using a calculator.

There are also properties of numbers that are true for certain operations. The *commutative* property allows the order of the terms in an expression to change while keeping the same final answer. Both addition and multiplication can be completed in any order and still obtain the same result. However, order does matter in subtraction and division. The *associative* property allows any terms to be "associated" by parenthesis and retain the same final answer. For example:

$$(4 + 3) + 5 = 4 + (3 + 5)$$

Both addition and multiplication are associative; however, subtraction and division do not hold this property. The *distributive* property states that:

$$a(b + c) = ab + ac$$

It is a property that involves both addition and multiplication, and the a is distributed onto each term inside the parentheses.

Integers can be factored into prime numbers. To *factor* is to express as a product. For example:

$$6 = 3 \cdot 2$$

and

$$6 = 6 \cdot 1$$

Both are factorizations, but the expression involving the factors of 3 and 2 is known as a *prime factorization* because it is factored into a product of two *prime numbers*—integers which do not have any factors other than themselves and 1. A *composite number* is a positive integer that can be divided into at least one other integer other than itself and 1, such as 6. Integers that have a factor of 2 are even, and if they are not divisible by 2, they are odd. Finally, a *multiple* of a number is the product of that number and a counting number—also known as a *natural number*. For example, some multiples of 4 are 4, 8, 12, 16, etc.

Addition with Whole Numbers and Fractions

Addition combines two quantities together. With whole numbers, this is taking two sets of things and merging them into one, then counting the result. For example, 4 + 3 = 7. When adding numbers, the order does not matter: 3 + 4 = 7, also. Longer lists of whole numbers can also be added together. The result of adding numbers is called the *sum*.

With fractions, the number on top is the *numerator*, and the number on the bottom is the *denominator*. To add fractions, the denominator must be the same—a *common denominator*. To find a common denominator, the existing numbers on the bottom must be considered, and the lowest number they will both multiply into must be determined. Consider the following equation:

$$\frac{1}{3} + \frac{5}{6} = ?$$

The numbers 3 and 6 both multiply into 6. Three can be multiplied by 2, and 6 can be multiplied by 1. The top and bottom of each fraction must be multiplied by the same number. Then, the numerators are added together to get a new numerator. The following equation is the result:

$$\frac{1}{3} + \frac{5}{6} = \frac{2}{6} + \frac{5}{6} = \frac{7}{6}$$

Subtraction with Whole Numbers and Fractions

Subtraction is taking one quantity away from another, so it is the opposite of addition. The expression 4 − 3 means taking 3 away from 4. So, 4 − 3 = 1. In this case, the order matters, since it entails taking one quantity away from the other, rather than just putting two quantities together. The result of subtraction is also called the *difference*.

To subtract fractions, the denominator must be the same. Then, subtract the numerators together to get a new numerator. Here is an example:

$$\frac{1}{3} - \frac{5}{6} = \frac{2}{6} - \frac{5}{6} = \frac{-3}{6} = -\frac{1}{2}$$

Multiplication with Whole Numbers and Fractions

Multiplication is a kind of repeated addition. The expression 4 × 5 is taking four sets, each of them having five things in them, and putting them all together. That means:

$$4 \times 5 = 5 + 5 + 5 + 5 = 20$$

As with addition, the order of the numbers does not matter. The result of a multiplication problem is called the *product*.

To multiply fractions, the numerators are multiplied to get the new numerator, and the denominators are multiplied to get the new denominator:

$$\frac{1}{3} \times \frac{5}{6} = \frac{1 \times 5}{3 \times 6} = \frac{5}{18}$$

When multiplying fractions, common factors can *cancel* or *divide into one another*, when factors that appear in the numerator of one fraction and the denominator of the other fraction. Here is an example:

$$\frac{1}{3} \times \frac{9}{8} = \frac{1}{1} \times \frac{3}{8} = 1 \times \frac{3}{8} = \frac{3}{8}$$

The numbers 3 and 9 have a common factor of 3, so that factor can be divided out.

Division with Whole Numbers and Fractions

Division is the opposite of multiplication. With whole numbers, it means splitting up one number into sets of equal size. For example, $16 \div 8$ is the number of sets of eight things that can be made out of sixteen things. Thus, $16 \div 8 = 2$. As with subtraction, the order of the numbers will make a difference, here. The answer to a division problem is called the *quotient*, while the number in front of the division sign is called the *dividend*, and the number behind the division sign is called the *divisor*.

To divide fractions, the first fraction must be multiplied with the reciprocal of the second fraction. The *reciprocal* of the fraction $\frac{x}{y}$ is the fraction $\frac{y}{x}$. Here is an example:

$$\frac{1}{3} \div \frac{5}{6} = \frac{1}{3} \times \frac{6}{5} = \frac{6}{15} = \frac{2}{5}$$

Factorization

Factors are the numbers multiplied to achieve a product. Thus, every product in a multiplication equation has, at minimum, two factors. Of course, some products will have more than two factors. For the sake of most discussions, assume that factors are positive integers.

To find a number's factors, start with 1 and the number itself. Then divide the number by 2, 3, 4, and so on, seeing if any divisors can divide the number without a remainder, keeping a list of those that do. Stop upon reaching either the number itself or another factor.

Let's find the factors of 45. Start with 1 and 45. Then try to divide 45 by 2, which fails. Now divide 45 by 3. The answer is 15, so 3 and 15 are now factors. Dividing by 4 doesn't work, and dividing by 5 leaves 9. Lastly, dividing 45 by 6, 7, and 8 all don't work. The next integer to try is 9, but this is already known to be a factor, so the factorization is complete. The factors of 45 are 1, 3, 5, 9, 15 and 45.

Prime Factorization

Prime factorization involves an additional step after breaking a number down to its factors: breaking down the factors until they are all prime numbers. A prime number is any number that can only be divided by 1 and itself. The prime numbers between 1 and 20 are 2, 3, 5, 7, 11, 13, 17, and 19. As a simple test, numbers that are even or end in 5 are not prime.

Let's break 129 down into its prime factors. First, the factors are 3 and 43. Both 3 and 43 are prime numbers, so we're done. But if 43 was not a prime number, then it would also need to be factorized until all of the factors are expressed as prime numbers.

Common Factor

A common factor is a factor shared by two numbers. Let's take 45 and 30 and find the common factors:

The factors of 45 are: 1, 3, 5, 9, 15, and 45.
The factors of 30 are: 1, 2, 3, 5, 6, 10, 15, and 30.
The common factors are 1, 3, 5, and 15.

Greatest Common Factor

The greatest common factor is the largest number among the shared, common factors. From the factors of 45 and 30, the common factors are 3, 5, and 15. Thus, 15 is the greatest common factor, as it's the largest number.

Least Common Multiple

The least common multiple is the smallest number that's a multiple of two numbers. Let's try to find the least common multiple of 4 and 9. The multiples of 4 are 4, 8, 12, 16, 20, 24, 28, 32, 36, and so on. For 9, the multiples are 9, 18, 27, 36, 45, 54, etc. Thus, the least common multiple of 4 and 9 is 36, the lowest number where 4 and 9 share multiples.

If two numbers share no factors besides 1 in common, then their least common multiple will be simply their product. If two numbers have common factors, then their least common multiple will be their product divided by their greatest common factor. This can be visualized by the formula $LCM = \frac{x \times y}{GCF}$, where x and y are some integers and LCM and GCF are their least common multiple and greatest common factor, respectively.

Fractions

A fraction is an equation that represents a part of a whole, but can also be used to present ratios or division problems. An example of a fraction is $\frac{x}{y}$. In this example, x is called the numerator, while y is the denominator. The numerator represents the number of parts, and the denominator is the total number of parts. They are separated by a line or slash, known as a fraction bar.. In simple fractions, the numerator and denominator can be nearly any integer. However, the denominator of a fraction can never be zero because dividing by zero is a function which is undefined.

Imagine that an apple pie has been baked for a holiday party, and the full pie has eight slices. After the party, there are five slices left. How could the amount of the pie that remains be expressed as a fraction? The numerator is 5 since there are 5 pieces left, and the denominator is 8 since there were eight total slices in the whole pie. Thus, expressed as a fraction, the leftover pie totals $\frac{5}{8}$ of the original amount.

Fractions come in three different varieties: proper fractions, improper fractions, and mixed numbers. Proper fractions have a numerator less than the denominator, such as $\frac{3}{8}$, but improper fractions have a numerator greater than the denominator, such as $\frac{15}{8}$. Mixed numbers combine a whole number with a proper fraction, such as $3\frac{1}{2}$. Any mixed number can be written as an improper fraction by multiplying the integer by the denominator, adding the product to the value of the numerator, and dividing the sum by the original denominator. For example:

$$3\frac{1}{2} = \frac{3 \times 2 + 1}{2} = \frac{7}{2}$$

Whole numbers can also be converted into fractions by placing the whole number as the numerator and making the denominator 1. For example, $3 = \frac{3}{1}$.

One of the most fundamental concepts of fractions is their ability to be manipulated by multiplication or division. This is possible since $\frac{n}{n}$ = 1 for any non-zero integer. As a result, multiplying or dividing by $\frac{n}{n}$ will not alter the original fraction since any number multiplied or divided by 1 doesn't change the value of that number. Fractions of the same value are known as equivalent fractions. For example, $\frac{2}{4}, \frac{4}{8}, \frac{50}{100}$, and $\frac{75}{150}$ are equivalent, as they all equal $\frac{1}{2}$.

Although many equivalent fractions exist, they are easier to compare and interpret when reduced or simplified. The numerator and denominator of a simple fraction will have no factors in common other than 1. When reducing or simplifying fractions, divide the numerator and denominator by the greatest common factor. A simple strategy is to divide the numerator and denominator by low numbers, like 2, 3, or 5 until arriving at a simple fraction, but the same thing could be achieved by determining the greatest common factor for both the numerator and denominator and dividing each by it. Using the first method is preferable when both the numerator and denominator are even, end in 5, or are obviously a multiple of another number. However, if no numbers seem to work, it will be necessary to factor the numerator and denominator to find the GCF. Let's look at examples:

1) Simplify the fraction $\frac{6}{8}$:

Dividing the numerator and denominator by 2 results in $\frac{3}{4}$, which is a simple fraction.

2) Simplify the fraction $\frac{12}{36}$:

Dividing the numerator and denominator by 2 leaves $\frac{6}{18}$. This isn't a simple fraction, as both the numerator and denominator have factors in common. Dividing each by 3 results in $\frac{2}{6}$, but this can be further simplified by dividing by 2 to get $\frac{1}{3}$. This is the simplest fraction, as the numerator is 1. In cases like this, multiple division operations can be avoided by determining the greatest common factor between the numerator and denominator.

3) Simplify the fraction $\frac{18}{54}$ by dividing by the greatest common factor:

First, determine the factors for the numerator and denominator. The factors of 18 are 1, 2, 3, 6, 9, and 18. The factors of 54 are 1, 2, 3, 6, 9, 18, 27, and 54. Thus, the greatest common factor is 18. Dividing $\frac{18}{54}$ by 18 leaves $\frac{1}{3}$, which is the simplest fraction. This method takes slightly more work, but it definitively arrives at the simplest fraction.

Operations with Fractions

Of the four basic operations that can be performed on fractions, the one which involves the least amount of work is multiplication. To multiply two fractions, simply multiply the numerators, multiply the denominators, and place the products as a fraction. Whole numbers and mixed numbers can also be expressed as a fraction, as described above, to multiply with a fraction. Let's work through a couple of examples.

1) $\frac{2}{5} \times \frac{3}{4} = \frac{6}{20} = \frac{3}{10}$

2) $\frac{4}{9} \times \frac{7}{11} = \frac{28}{99}$

Dividing fractions is similar to multiplication with one key difference. To divide fractions, flip the numerator and denominator of the second fraction, and then proceed as if it were a multiplication problem:

1) $\frac{7}{8} \div \frac{4}{5} = \frac{7}{8} \times \frac{5}{4} = \frac{35}{32}$

2) $\frac{5}{9} \div \frac{1}{3} = \frac{5}{9} \times \frac{3}{1} = \frac{15}{9} = \frac{5}{3}$

Addition and subtraction require more steps than multiplication and division, as these operations require the fractions to have the same denominator, also called a common denominator. It is always possible to find a common denominator by multiplying the denominators. However, when the denominators are large numbers, this method is unwieldy, especially if the answer must be provided in its simplest form. Thus, it's beneficial to find the least common denominator of the fractions—the least common denominator is incidentally also the least common multiple.

Once equivalent fractions have been found with common denominators, simply add or subtract the numerators to arrive at the answer:

1) $\frac{1}{2} + \frac{3}{4} = \frac{2}{4} + \frac{3}{4} = \frac{5}{4}$

2) $\frac{3}{12} + \frac{11}{20} = \frac{15}{60} + \frac{33}{60} = \frac{48}{60} = \frac{4}{5}$

3) $\frac{7}{9} - \frac{4}{15} = \frac{35}{45} - \frac{12}{45} = \frac{23}{45}$

4) $\frac{5}{6} - \frac{7}{18} = \frac{15}{18} - \frac{7}{18} = \frac{8}{18} = \frac{4}{9}$

Recognizing Equivalent Fractions and Mixed Numbers

The value of a fraction does not change if multiplying or dividing both the numerator and the denominator by the same number (other than 0). In other words:

$$\frac{x}{y} = \frac{a \times x}{a \times y} = \frac{x \div a}{y \div a}$$

as long as a is not 0. This means that $\frac{2}{5} = \frac{4}{10}$, for example. If x and y are integers that have no common factors, then the fraction is said to be *simplified*. This means $\frac{2}{5}$ is simplified, but $\frac{4}{10}$ is not.

Often when working with fractions, the fractions need to be rewritten so that they all share a single denominator—this is called finding a *common denominator* for the fractions. Using two fractions, $\frac{a}{b}$ and $\frac{c}{d}$, the numerator and denominator of the left fraction can be multiplied by d, while the numerator and denominator of the right fraction can be multiplied by b. This provides the fractions $\frac{a \times d}{b \times d}$ and $\frac{c \times b}{d \times b}$ with the common denominator $b \times d$.

A fraction whose numerator is smaller than its denominator is called a *proper fraction*. A fraction whose numerator is bigger than its denominator is called an *improper fraction*. These numbers can be rewritten as a combination of integers and fractions, called a *mixed number*. For example:

$$\frac{6}{5} = \frac{5}{5} + \frac{1}{5} = 1 + \frac{1}{5}$$

and can be written as $1\frac{1}{5}$.

Distribution of a Quantity into its Fractional Parts

A quantity may be broken into its fractional parts. For example, a toy box holds three types of toys for kids. $\frac{1}{3}$ of the toys are Type A and $\frac{1}{4}$ of the toys are Type B. With that information, how many Type C toys are there?

First, the sum of Type A and Type B must be determined by finding a common denominator to add the fractions. The lowest common multiple is 12, so that is what will be used. The sum is:

$$\frac{1}{3} + \frac{1}{4} = \frac{4}{12} + \frac{3}{12} = \frac{7}{12}$$

This value is subtracted from 1 to find the number of Type C toys. The value is subtracted from 1 because 1 represents a whole. The calculation is:

$$1 - \frac{7}{12} = \frac{12}{12} - \frac{7}{12} = \frac{5}{12}$$

This means that $\frac{5}{12}$ of the toys are Type C. To check the answer, add all fractions together, and the result should be 1.

Recognition of Decimals

The *decimal system* is a way of writing out numbers that uses ten different numerals: 0, 1, 2, 3, 4, 5, 6, 7, 8, and 9. This is also called a "base ten" or "base 10" system. Other bases are also used. For example, computers work with a base of 2. This means they only use the numerals 0 and 1.

The *decimal place* denotes how far to the right of the decimal point a numeral is. The first digit to the right of the decimal point is in the *tenths* place. The next is the *hundredths*. The third is the *thousandths*.

So, 3.142 has a 1 in the tenths place, a 4 in the hundredths place, and a 2 in the thousandths place.

The *decimal point* is a period used to separate the *ones* place from the *tenths* place when writing out a number as a decimal.

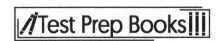

A *decimal number* is a number written out with a decimal point instead of as a fraction, for example, 1.25 instead of $\frac{5}{4}$. Depending on the situation, it can sometimes be easier to work with fractions and sometimes easier to work with decimal numbers.

A decimal number is *terminating* if it stops at some point. It is called *repeating* if it never stops, but repeats a pattern over and over. It is important to note that every rational number can be written as a terminating decimal or as a repeating decimal.

Addition with Decimals

To add decimal numbers, each number in columns needs to be lined up by the decimal point. For each number being added, the zeros to the right of the last number need to be filled in so that each of the numbers has the same number of places to the right of the decimal. Then, the columns can be added together. Here is an example of 2.45 + 1.3 + 8.891 written in column form:

$$\begin{array}{r} 2.450 \\ 1.300 \\ + \,8.891 \end{array}$$

Zeros have been added in the columns so that each number has the same number of places to the right of the decimal.

Added together, the correct answer is 12.641:

$$\begin{array}{r} 2.450 \\ 1.300 \\ + \,8.891 \\ \hline 12.641 \end{array}$$

Subtraction with Decimals

Subtracting decimal numbers is the same process as adding decimals. Here is 7.89 − 4.235 written in column form:

$$\begin{array}{r} 7.890 \\ - \,4.235 \\ \hline 3.655 \end{array}$$

A zero has been added in the column so that each number has the same number of places to the right of the decimal.

Multiplication with Decimals

The simplest way to multiply decimals is to calculate the product as if the decimals are not there, then count the number of decimal places in the original problem. Use that total to place the decimal the same number of places over in your answer, counting from right to left. For example, 0.5 x 1.25 can be rewritten and multiplied as 5 x 125, which equals 625. Then the decimal is added three places from the right for .625.

The final answer will have the same number of decimal *points* as the total number of decimal *places* in the problem. The first number has one decimal place, and the second number has two decimal places. Therefore, the final answer will contain three decimal places:

$$0.5 \times 1.25 = 0.625$$

Division with Decimals

Dividing a decimal by a whole number entails using long division first by ignoring the decimal point. Then, the decimal point is moved the number of places given in the problem.

For example, $6.8 \div 4$ can be rewritten as $68 \div 4$, which is 17. There is one non-zero integer to the right of the decimal point, so the final solution would have one decimal place to the right of the solution. In this case, the solution is 1.7.

Dividing a decimal by another decimal requires changing the divisor to a whole number by moving its decimal point. The decimal place of the dividend should be moved by the same number of places as the divisor. Then, the problem is the same as dividing a decimal by a whole number.

For example, $5.72 \div 1.1$ has a divisor with one decimal point in the denominator. The expression can be rewritten as $57.2 \div 11$ by moving each number one decimal place to the right to eliminate the decimal. The long division can be completed as $572 \div 11$ with a result of 52. Since there is one non-zero integer to the right of the decimal point in the problem, the final solution is 5.2.

In another example, $8 \div 0.16$ has a divisor with two decimal points in the denominator. The expression can be rewritten as $800 \div 16$ by moving each number two decimal places to the right to eliminate the decimal in the divisor. The long division can be completed with a result of 50.

Percentages

Think of percentages as fractions with a denominator of 100. In fact, percentage means "per hundred." Problems often require converting numbers from percentages, fractions, and decimals.

Percent Problems

The basic percent equation is the following:

$$\frac{is}{of} = \frac{\%}{100}$$

The placement of numbers in the equation depends on what the question asks.

Example 1
Find 40% of 80.

Basically, the problem is asking, "What is 40% of 80?" The 40% is the percent, and 80 is the number to find the percent "of." The equation is:

$$\frac{x}{80} = \frac{40}{100}$$

Solving the equation by cross-multiplication, the problem becomes 100x = 80(40). Solving for x gives the answer: x = 32.

Example 2
What percent of 100 is 20?

The 20 fills in the "is" portion, while 100 fills in the "of." The question asks for the percent, so that will be x, the unknown. The following equation is set up:

$$\frac{20}{100} = \frac{x}{100}$$

Cross-multiplying yields the equation $100x = 20(100)$. Solving for x gives the answer of 20%.

Example 3
30% of what number is 30?

The following equation uses the clues and numbers in the problem:

$$\frac{30}{x} = \frac{30}{100}$$

Cross-multiplying results in the equation $30(100) = 30x$. Solving for x gives the answer $x = 100$.

Fraction and Percent Equivalencies

The word *percent* comes from the Latin phrase for "per one hundred." A *percent* is a way of writing out a fraction. It is a fraction with a denominator of 100. Thus, $65\% = \frac{65}{100}$.

To convert a fraction to a percent, the denominator is written as 100. For example, $\frac{3}{5} = \frac{60}{100} = 60\%$.

In converting a percent to a fraction, the percent is written with a denominator of 100, and the result is simplified. For example, $30\% = \frac{30}{100} = \frac{3}{10}$.

Conversions

Decimals and Percentages
Since a percentage is based on "per hundred," decimals and percentages can be converted by multiplying or dividing by 100. Practically speaking, this always amounts to moving the decimal point two places to the right or left, depending on the conversion. To convert a percentage to a decimal, move the decimal point two places to the left and remove the % sign. To convert a decimal to a percentage, move the decimal point two places to the right and add a "%" sign. Here are some examples:

65% = 0.65
0.33 = 33%
0.215 = 21.5%
99.99% = 0.9999
500% = 5.00
7.55 = 755%

Fractions and Percentages

Remember that a percentage is a number per one hundred. So a percentage can be converted to a fraction by making the number in the percentage the numerator and putting 100 as the denominator:

$$43\% = \frac{43}{100}$$

$$97\% = \frac{97}{100}$$

Note that the percent symbol (%) kind of looks like a 0, a 1, and another 0. So think of a percentage like 54% as 54 over 100.

To convert a fraction to a percent, follow the same logic. If the fraction happens to have 100 in the denominator, you're in luck. Just take the numerator and add a percent symbol:

$$\frac{28}{100} = 28\%$$

Otherwise, divide the numerator by the denominator to get a decimal:

$$\frac{9}{12} = 0.75$$

Then convert the decimal to a percentage:

$$0.75 = 75\%$$

Another option is to make the denominator equal to 100. Be sure to multiply the numerator by the same number as the denominator. For example:

$$\frac{3}{20} \times \frac{5}{5} = \frac{15}{100}$$

$$\frac{15}{100} = 15\%$$

Changing Fractions to Decimals

To change a fraction into a decimal, divide the denominator into the numerator until there are no remainders. There may be repeating decimals, so rounding is often acceptable. A straight line above the repeating portion denotes that the decimal repeats.

Example

Express 4/5 as a decimal.

Set up the division problem.

$$5\overline{)4}$$

5 does not go into 4, so place the decimal and add a zero.

$$5\overline{)4.0}$$

5 goes into 40 eight times. There is no remainder.

$$
\begin{array}{r}
0.8 \\
5\overline{)4.0} \\
-4.0 \\
\hline
0
\end{array}
$$

The solution is 0.8.

Example

Express 33 1/3 as a decimal.

Since the whole portion of the number is known, set it aside to calculate the decimal from the fraction portion.

Set up the division problem.

$$3\overline{)1}$$

3 does not go into 1, so place the decimal and add zeros. 3 goes into 10 three times.

$$
\begin{array}{r}
0.333 \\
3\overline{)1.000}
\end{array}
$$

This will repeat with a remainder of 1, so place a line over the 3 denotes the repetition.

$$
\begin{array}{r}
0.333 \\
3\overline{)1.000} \\
-9 \\
\hline
10 \\
-9 \\
\hline
10
\end{array}
$$

The solution is $0.\overline{3}$

Changing Decimals to Fractions

To change decimals to fractions, place the decimal portion of the number, the numerator, over the respective place value, the denominator, then reduce, if possible.

Example

Express 0.25 as a fraction.

This is read as twenty-five hundredths, so put 25 over 100. Then reduce to find the solution.

$$\frac{25}{100} = \frac{1}{4}$$

Example

Express 0.455 as a fraction

This is read as four hundred fifty-five thousandths, so put 455 over 1000. Then reduce to find the solution.

$$\frac{455}{1000} = \frac{91}{200}$$

There are two types of problems that commonly involve percentages. The first is to calculate some percentage of a given quantity, where you convert the percentage to a decimal, and multiply the quantity by that decimal. Secondly, you are given a quantity and told it is a fixed percent of an unknown quantity. In this case, convert to a decimal, then divide the given quantity by that decimal.

Example

What is 30% of 760?

Convert the percent into a useable number. "Of" means to multiply.

$$30\% = 0.30$$

Set up the problem based on the givens, and solve.

$$0.30 \times 760 = 228$$

Example

8.4 is 20% of what number?

Convert the percent into a useable number.

$$20\% = 0.20$$

The given number is a percent of the answer needed, so divide the given number by this decimal rather than multiplying it.

$$\frac{8.4}{0.20} = 42$$

Word Problems

Word problems can appear daunting, but don't let the verbiage psyche you out. No matter the scenario or specifics, the key to answering them is to translate the words into a math problem. Always keep in mind what the question is asking and what operations could lead to that answer. The following word problems highlight the most commonly tested question types.

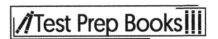

Working with Money

Walter's Coffee Shop sells a variety of drinks and breakfast treats.

Price List	
Hot Coffee	$2.00
Slow-Drip Iced Coffee	$3.00
Latte	$4.00
Muffin	$2.00
Crepe	$4.00
Egg Sandwich	$5.00

Costs	
Hot Coffee	$0.25
Slow-Drip Iced Coffee	$0.75
Latte	$1.00
Muffin	$1.00
Crepe	$2.00
Egg Sandwich	$3.00

Walter's utilities, rent, and labor costs him $500 per day. Today, Walter sold 200 hot coffees, 100 slow-drip iced coffees, 50 lattes, 75 muffins, 45 crepes, and 60 egg sandwiches. What was Walter's total profit today?

To accurately answer this type of question, determine the total cost of making his drinks and treats, then determine how much revenue he earned from selling those products. After arriving at these two totals, the profit is measured by deducting the total cost from the total revenue.

Walter's costs for today:

Item	Quantity	Cost Per Unit	Total Cost
Hot Coffee	200	$0.25	$50
Slow-Drip Iced Coffee	100	$0.75	$75
Latte	50	$1.00	$50
Muffin	75	$1.00	$75
Crepe	45	$2.00	$90
Egg Sandwich	60	$3.00	$180
Utilities, rent, and labor			$500
Total Costs			$1,020

Walter's revenue for today:

Item	Quantity	Revenue Per Unit	Total Revenue
Hot Coffee	200	$2.00	$400
Slow-Drip Iced Coffee	100	$3.00	$300
Latte	50	$4.00	$200
Muffin	75	$2.00	$150
Crepe	45	$4.00	$180
Egg Sandwich	60	$5.00	$300
Total Revenue			$1,530

Walter's Profit = *Revenue − Costs* = $1,530 − $1,020 = $510

This strategy is applicable to other question types. For example, calculating salary after deductions, balancing a checkbook, and calculating a dinner bill are common word problems similar to business planning. Just remember to use the correct operations. When a balance is increased, use addition. When a balance is decreased, use subtraction. Common sense and organization are your greatest assets when answering word problems.

Solving Real-World Problems Involving Percentages

Questions dealing with percentages can be difficult when they are phrased as word problems. These word problems almost always come in three varieties. The first type will ask to find what percentage of some number will equal another number. The second asks to determine what number is some percentage of another given number. The third will ask what number another number is a given percentage of.

One of the most important parts of correctly answering percentage word problems is to identify the numerator and the denominator. This fraction can then be converted into a percentage, as described above.

The following word problem shows how to make this conversion:

A department store carries several different types of footwear. The store is currently selling 8 athletic shoes, 7 dress shoes, and 5 sandals. What percentage of the store's footwear are sandals?

First, calculate what serves as the 'whole', as this will be the denominator. How many total pieces of footwear does the store sell? The store sells 20 different types (8 athletic + 7 dress + 5 sandals).

Second, what footwear type is the question specifically asking about? Sandals. Thus, 5 is the numerator.

Third, the resultant fraction must be expressed as a percentage. The first two steps indicate that $\frac{5}{20}$ of the footwear pieces are sandals. This fraction must now be converted into a percentage:

$$\frac{5}{20} \times \frac{5}{5} = \frac{25}{100} = 25\%$$

Solving Real-World Problems Involving Proportions

Much like a scale factor can be written using an equation like $2A = B$, a *relationship* is represented by the equation $Y = kX$. X and Y are proportional because as values of X increase, the values of Y also increase. A relationship that is inversely proportional can be represented by the equation $Y = \frac{k}{X}$, where the value of Y decreases as the value of x increases and vice versa.

Proportional reasoning can be used to solve problems involving ratios, percentages, and averages. Ratios can be used in setting up proportions and solving them to find unknowns. For example, if student completes an average of 10 pages of math homework in 3 nights, how long would it take the student to complete 22 pages? Both ratios can be written as fractions. The second ratio would contain the unknown.

The following proportion represents this problem, where x is the unknown number of nights:

$$\frac{10 \; pages}{3 \; nights} = \frac{22 \; pages}{x \; nights}$$

Solving this proportion entails cross-multiplying and results in the following equation: $10x = 22 * 3$. Simplifying and solving for x results in the exact solution: $x = 6.6 \; nights$. The result would be rounded up to 7 because the homework would actually be completed on the 7th night.

The following problem uses ratios involving percentages:

If 20% of the class is girls and 30 students are in the class, how many girls are in the class?

To set up this problem, it is helpful to use the common proportion:

$$\frac{\%}{100} = \frac{is}{of}$$

Within the proportion, % is the percentage of girls, 100 is the total percentage of the class, *is* is the number of girls, and *of* is the total number of students in the class. Most percentage problems can be written using this language. To solve this problem, the proportion should be set up as

$$\frac{20}{100} = \frac{x}{30}$$

and then solved for x. Cross-multiplying results in the equation $20 * 30 = 100x$, which results in the solution $x = 6$. There are 6 girls in the class.

Problems involving volume, length, and other units can also be solved using ratios. For example, a problem may ask for the volume of a cone to be found that has a radius, $r = 7m$ and a height, $h = 16m$. Referring to the formulas provided on the test, the volume of a cone is given as:

$$V = \pi r^2 \frac{h}{3}$$

where r is the radius, and h is the height. Plugging $r = 7$ and $h = 16$ into the formula, the following is obtained:

$$V = \pi (7^2) \frac{16}{3}$$

Therefore, volume of the cone is found to be approximately 821m³. Sometimes, answers in different units are sought. If this problem wanted the answer in liters, 821m³ would need to be converted. Using the equivalence statement 1m³ = 1000L, the following ratio would be used to solve for liters: 821m³ ∗ $\frac{1000L}{1m^3}$. Cubic meters in the numerator and denominator cancel each other out, and the answer is converted to 821,000 liters, or $8.21 * 10^5$ L.

Other conversions can also be made between different given and final units. If the temperature in a pool is 30°C, what is the temperature of the pool in degrees Fahrenheit? To convert these units, an equation is used relating Celsius to Fahrenheit. The following equation is used:

$$T_{^\circ F} = 1.8 T_{^\circ C} + 32$$

Plugging in the given temperature and solving the equation for T yields the result:

$$T_{^\circ F} = 1.8(30) + 32 = 86^\circ F$$

Both units in the metric system and U.S. customary system are widely used.

Here are some more examples of how to solve for proportions:

1) $\frac{75\%}{90\%} = \frac{25\%}{x}$

To solve for x, the fractions must be cross multiplied:

$$(75\%x = 90\% \times 25\%)$$

To make things easier, let's convert the percentages to decimals:

$$(0.9 \times 0.25 = 0.225 = 0.75x)$$

To get rid of x's co-efficient, each side must be divided by that same coefficient to get the answer $x = 0.3$. The question could ask for the answer as a percentage or fraction in lowest terms, which are 30% and $\frac{3}{10}$, respectively.

2) $\frac{x}{12} = \frac{30}{96}$

Cross-multiply: $96x = 30 \times 12$

Multiply: $96x = 360$

Divide: $x = 360 \div 96$

Answer: $x = 3.75$

3) $\frac{0.5}{3} = \frac{x}{6}$

Cross-multiply: $3x = 0.5 \times 6$

Multiply: $3x = 3$

Divide: $x = 3 \div 3$

Answer: $x = 1$

You may have noticed there's a faster way to arrive at the answer. If there is an obvious operation being performed on the proportion, the same operation can be used on the other side of the proportion to solve for x. For example, in the first practice problem, 75% became 25% when divided by 3, and upon doing the same to 90%, the correct answer of 30% would have been found with much less legwork. However, these questions aren't always so intuitive, so it's a good idea to work through the steps, even if the answer seems apparent from the outset.

Solving Real-World Problems Involving Ratios and Rates of Change

Ratios are used to show the relationship between two quantities. The ratio of oranges to apples in the grocery store may be 3 to 2. That means that for every 3 oranges, there are 2 apples. This comparison can be expanded to represent the actual number of oranges and apples. Another example may be the number of boys to girls in a math class. If the ratio of boys to girls is given as 2 to 5, that means there are 2 boys to every 5 girls in the class. Ratios can also be compared if the units in each ratio are the same. The ratio of boys to girls in the math class can be compared to the ratio of boys to girls in a science class by stating which ratio is higher and which is lower.

Rates are used to compare two quantities with different units. *Unit rates* are the simplest form of rate. With unit rates, the denominator in the comparison of two units is one. For example, if someone can type at a rate of 1000 words in 5 minutes, then his or her unit rate for typing is $\frac{1000}{5} = 200$ words in one minute or 200 words per minute. Any rate can be converted into a unit rate by dividing to make the denominator one. 1000 words in 5 minutes has been converted into the unit rate of 200 words per minute.

Ratios and rates can be used together to convert rates into different units. For example, if someone is driving 50 kilometers per hour, that rate can be converted into miles per hour by using a ratio known as the *conversion factor*. Since the given value contains kilometers and the final answer needs to be in miles, the ratio relating miles to kilometers needs to be used. There are 0.62 miles in 1 kilometer. This, written as a ratio and in fraction form, is $\frac{0.62 \ miles}{1 \ km}$. To convert 50km/hour into miles per hour, the following conversion needs to be set up:

$$\frac{50 \ km}{hour} * \frac{0.62 \ miles}{1 \ km} = 31 \ miles \ per \ hour$$

The ratio between two similar geometric figures is called the *scale factor*. For example, a problem may depict two similar triangles, A and B. The scale factor from the smaller triangle A to the larger triangle B is given as 2 because the length of the corresponding side of the larger triangle, 16, is twice the corresponding side on the smaller triangle, 8. This scale factor can also be used to find the value of a missing side, x, in triangle A. Since the scale factor from the smaller triangle (A) to larger one (B) is 2, the larger corresponding side in triangle B (given as 25), can be divided by 2 to find the missing side in A (x = 12.5). The scale factor can also be represented in the equation $2A = B$ because two times the lengths of A gives the corresponding lengths of B. This is the idea behind similar triangles.

Unit Rate

Unit rate word problems will ask to calculate the rate or quantity of something in a different value. For example, a problem might say that a car drove a certain number of miles in a certain number of minutes and then ask how many miles per hour the car was traveling. These questions involve solving proportions. Consider the following examples:

1) Alexandra made $96 during the first 3 hours of her shift as a temporary worker at a law office. She will continue to earn money at this rate until she finishes in 5 more hours. How much does Alexandra make per hour? How much will Alexandra have made at the end of the day?

This problem can be solved in two ways. The first is to set up a proportion, as the rate of pay is constant. The second is to determine her hourly rate, multiply the 5 hours by that rate, and then add the $96.

To set up a proportion, put the money already earned over the hours already worked on one side of an equation. The other side has x over 8 hours (the total hours worked in the day). It looks like this: $\frac{96}{3} = \frac{x}{8}$. Now, cross-multiply to get $768 = 3x$. To get x, divide by 3, which leaves $x = 256$. Alternatively, as x is the numerator of one of the proportions, multiplying by its denominator will reduce the solution by one step. Thus, Alexandra will make $256 at the end of the day. To calculate her hourly rate, divide the total by 8, giving $32 per hour.

Alternatively, it is possible to figure out the hourly rate by dividing $96 by 3 hours to get $32 per hour. Now her total pay can be figured by multiplying $32 per hour by 8 hours, which comes out to $256.

2) Jonathan is reading a novel. So far, he has read 215 of the 335 total pages. It takes Jonathan 25 minutes to read 10 pages, and the rate is constant. How long does it take Jonathan to read one page? How much longer will it take him to finish the novel? Express the answer in time.

To calculate how long it takes Jonathan to read one page, divide the 25 minutes by 10 pages to determine the page per minute rate. Thus, it takes 2.5 minutes to read one page.

Jonathan must read 120 more pages to complete the novel. (This is calculated by subtracting the pages already read from the total.) Now, multiply his rate per page by the number of pages. Thus, $120 \times 2.5 = 300$. Expressed in time, 300 minutes is equal to 5 hours.

3) At a hotel, $\frac{4}{5}$ of the 120 rooms are booked for Saturday. On Sunday, $\frac{3}{4}$ of the rooms are booked. On which day are more of the rooms booked, and by how many more?

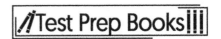

The first step is to calculate the number of rooms booked for each day. Do this by multiplying the fraction of the rooms booked by the total number of rooms.

Saturday: $\frac{4}{5} \times 120 = \frac{4}{5} \times \frac{120}{1} = \frac{480}{5} = 96$ rooms

Sunday: $\frac{3}{4} \times 120 = \frac{3}{4} \times \frac{120}{1} = \frac{360}{4} = 90$ rooms

Thus, more rooms were booked on Saturday by 6 rooms.

4) In a veterinary hospital, the veterinarian-to-pet ratio is 1:9. The ratio is always constant. If there are 45 pets in the hospital, how many veterinarians are currently in the veterinary hospital?

Set up a proportion to solve for the number of veterinarians: $\frac{1}{9} = \frac{x}{45}$

Cross-multiplying results in $9x = 45$, which works out to 5 veterinarians.

Alternatively, as there are always 9 times as many pets as veterinarians, is it possible to divide the number of pets (45) by 9. This also arrives at the correct answer of 5 veterinarians.

5) At a general practice law firm, 30% of the lawyers work solely on tort cases. If 9 lawyers work solely on tort cases, how many lawyers work at the firm?

First, solve for the total number of lawyers working at the firm, which will be represented here with x. The problem states that 9 lawyers work solely on torts cases, and they make up 30% of the total lawyers at the firm. Thus, 30% multiplied by the total, x, will equal 9. Written as equation, this is: $30\% \times x = 9$.

It's easier to deal with the equation after converting the percentage to a decimal, leaving $0.3x = 9$. Thus, $x = \frac{9}{0.3} = 30$ lawyers working at the firm.

6) Xavier was hospitalized with pneumonia. He was originally given 35mg of antibiotics. Later, after his condition continued to worsen, Xavier's dosage was increased to 60mg. What was the percent increase of the antibiotics? Round the percentage to the nearest tenth.

An increase or decrease in percentage can be calculated by dividing the difference in amounts by the original amount and multiplying by 100. Written as an equation, the formula is:

$$\frac{new\ quantity - old\ quantity}{old\ quantity} \times 100$$

Here, the question states that the dosage was increased from 35mg to 60mg, so these are plugged into the formula to find the percentage increase.

$$\frac{60 - 35}{35} \times 100 = \frac{25}{35} \times 100 = .7142 \times 100 = 71.4\%$$

Rate of Change

Rate of change for any line calculates the steepness of the line over a given interval. Rate of change is also known as the slope or rise/run. The TEAS will focus on the rate of change for linear functions which are straight lines. The slope is given by the change in *y* divided by the change in *x*. So the formula looks like this:

$$slope = \frac{y_2 - y_1}{x_2 - x_1}$$

In the graph below, two points are plotted. The first has the coordinates of (0,1), and the second point is (2,3). Remember that the x coordinate is always placed first in coordinate pairs. Work from left to right when identifying coordinates. Thus, the point on the left is point 1 (0,1), and the point on the right is point 2 (2,3).

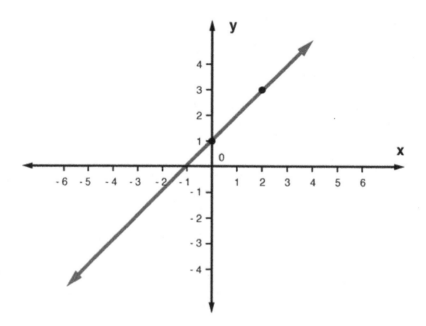

Now we need to just plug those numbers into the equation:

$$slope = \frac{3 - 1}{2 - 0}$$

$$slope = \frac{2}{2}$$

$$slope = 1$$

This means that for every increase of 1 for x, y also increased by 1. You can see this in the line. When x equalled 0, y equalled 1, and when x was increased to 1, y equalled 2.

Slope can be thought of as determining the rise over run:

$$slope = \frac{rise}{run}$$

The rise being the change vertically on the y axis and the run being the change horizontally on the x axis.

Translating Phrases and Sentences into Expressions, Equations, and Inequalities

To translate a word problem into an expression, look for a series of key words indicating addition, subtraction, multiplication, or division:

Addition: add, altogether, together, plus, increased by, more than, in all, sum, and total

Subtraction: minus, less than, difference, decreased by, fewer than, remain, and take away

Multiplication: *times*, *twice*, *of*, *double*, and *triple*

Division: divided by, cut up, half, quotient of, split, and shared equally

If a question asks to give words to a mathematical expression and says "equals," then an = sign must be included in the answer. Similarly, "less than or equal to" is expressed by the inequality symbol ≤, and "greater than or equal" to is expressed as ≥. Furthermore, "less than" is represented by <, and "greater than" is expressed by >.

Algebraic Patterns and Connections

Solving for X in Proportions

Proportions are commonly used to solve word problems to find unknown values such as x that are some percent or fraction of a known number. Proportions are solved by cross-multiplying and then dividing to arrive at x. The following examples show how this is done:

1) $\frac{75\%}{90\%} = \frac{25\%}{x}$

To solve for x, the fractions must be cross multiplied: ($75\%x = 90\% \times 25\%$). To make things easier, let's convert the percentages to decimals: ($0.9 \times 0.25 = 0.225 = 0.75x$). To get rid of x's co-efficient, each side must be divided by that same coefficient to get the answer $x = 0.3$. The question could ask for the answer as a percentage or fraction in lowest terms, which are 30% and $\frac{3}{10}$, respectively.

2) $\frac{x}{12} = \frac{30}{96}$

Cross-multiply: $96x = 30 \times 12$
Multiply: $96x = 360$
Divide: $x = 360 \div 96$
Answer: $x = 3.75$

3) $\frac{0.5}{3} = \frac{x}{6}$

Cross-multiply: $3x = 0.5 \times 6$
Multiply: $3x = 3$
Divide: $x = 3 \div 3$
Answer: $x = 1$

You may have noticed there's a faster way to arrive at the answer. If there is an obvious operation being performed on the proportion, the same operation can be used on the other side of the proportion to solve for x. For example, in the first practice problem, 75% became 25% when divided by 3, and upon doing the same to 90%, the correct answer of 30% would have been found with much less legwork. However, these questions aren't always so intuitive, so it's a good idea to work through the steps, even if the answer seems apparent from the outset.

There is a specific order in which operations are performed. This order is remembered using the acronym PEMDAS. PEMDAS stands for parenthesis, exponents, multiplication/division, and addition/subtraction. Multiplication and division are performed in the same step, working from left to right with whichever comes first. Addition and subtraction are performed in the same step, working from left to right with whichever comes first.

A memory device to help recall the order is word PEMDAS, or "Please Excuse My Dear Aunt Sally."

Example
Solve using correct order of operations $(3 + 4)(4 \div 2) + 8$.

Calculate anything inside parentheses first.

$$(3 + 4)(4 \div 2) + 8$$

Multiply.

$$(7)(2) + 8$$

Add, then solve.

$$14 + 8 = 22$$

Not every equation contains every operator, but the order of operations needs to be followed to obtain the correct answer. Some tell you the number that represents a variable. In that case, replace the variable with the number first, then follow the order to solve.

Example
Solve $X^2 + 5 - 1$, for $X = 3$.

Replace X with 3.

$$X^2 + 5 - 1$$

Use order of operations to solve exponents.

$$3^2 + 5 - 1$$

Add and subtract.

$$9 + 5 - 1$$

Solve.

$$14 - 1 = 13$$

Another application for algebra is temperature conversion. The United States commonly uses the Fahrenheit scale to measure temperature. In science and medicine, the Celsius scale is used. Here are some common comparisons:

$0 \, {}^{0}C \, = \, 32 \, {}^{0}F$, water freezes.

$100 \, {}^{0}C \, = \, 212 \, {}^{0}F$, water boils.

To convert between the two temperature scales, use the following equations.

$$9/5 \, {}^{0}C \, + \, 32 \, = \, {}^{0}F$$

$$5/9 \, ({}^{0}F \, - \, 32) \, = \, {}^{0}C$$

Example
A patient has a temperature of 39.6 °C. Convert this to Fahrenheit to assess whether she needs medical care.

Set up the equation using 39.6 for °C.

$$9/5 \, (39.6) \, + \, 32 \, = \, {}^{0}F$$

Multiply, divide and then add.

$$71.28 \, + \, 32 \, = \, {}^{0}F$$

Solve.

$$103.28 \, {}^{0}F$$

With a temperature over 103 °F, the patient may need medical care.

When trying to isolate a term or solve for a variable on one side of an equation, it is important not to change the equation. Always do the same operations to both sides of the equation.

Example
Solve for X, $X \, - \, 9 \, = \, 10$.

Solve for X by isolating it on a side.

$$X - 9 \, = \, 10$$

To get X alone, eliminate the 9 by adding 9 to both sides.

$$X - 9 + 9 \, = \, 10 + 9$$

Solve.

$$X \, = \, 19$$

Example
Solve for X, $4X = 20$.

Solve for X by isolating it on one side.

$$4X = 20$$

To get X alone, eliminate the 4 by dividing both sides by 4.

$$\frac{4x}{4} = \frac{20}{4}$$

Solve

$$X = 5$$

Example
Solve for X, $X^2 - 2 = 7$.

Isolate X on a side by adding 2 to both sides.

$$X^2 - 2 + 2 = 7 + 2$$

To undo the squaring of X, take the square root of both sides.

$$\sqrt{X^2} = \sqrt{9}$$

Solve.

$$X = 3$$

While solving an equation, you can also combine like terms. This is also called simplifying.

Examples of like terms would be X^2 and $3X^2$, or 4X and 8X.

Simplify: $X^2 + 2X^2 + 9X - 3X + 1 - 5$. This is not a full equation so we cannot solve it, only simplify it.

Identify all like terms.

$$X^2 + 2X^2 + 9X - 3X + 1 - 5$$

Combine the terms. Be sure to use the proper signs.

$$3X^2 + 6X - 4$$

Ratio Problems
A *ratio* compares the size of one group to the size of another. For example, there may be a room with 4 tables and 24 chairs. The ratio of tables to chairs is $4:24$. Such ratios behave like fractions in that both sides of the ratio by the same number can be multiplied or divided. Thus, the ratio 4:24 is the same as the ratio 2:12 and 1:6.

One quantity is *proportional* to another quantity if the first quantity is always some multiple of the second. For instance, the distance travelled in five hours is always five times to the speed as travelled. The distance is proportional to speed in this case.

One quantity is *inversely proportional* to another quantity if the first quantity is equal to some number divided by the second quantity. The time it takes to travel one hundred miles will be given by 100 divided by the speed travelled. The time is inversely proportional to the speed.

When dealing with word problems, there is no fixed series of steps to follow, but there are some general guidelines to use. It is important that the quantity to be found is identified. Then, it can be determined how the given values can be used and manipulated to find the final answer.

Example 1

Jana wants to travel to visit Alice, who lives one hundred and fifty miles away. If she can drive at fifty miles per hour, how long will her trip take?

The quantity to find is the *time* of the trip. The time of a trip is given by the distance to travel divided by the speed to be traveled. The problem determines that the distance is one hundred and fifty miles, while the speed is fifty miles per hour. Thus, 150 divided by 50 is $150 \div 50 = 3$. Because *miles* and *miles per hour* are the units being divided, the miles cancel out. The result is 3 hours.

Example 2

Bernard wishes to paint a wall that measures twenty feet wide by eight feet high. It costs ten cents to paint one square foot. How much money will Bernard need for paint?

The final quantity to compute is the *cost* to paint the wall. This will be ten cents ($0.10) for each square foot of area needed to paint. The area to be painted is unknown, but the dimensions of the wall are given; thus, it can be calculated.

The dimensions of the wall are 20 feet wide and 8 feet high. Since the area of a rectangle is length multiplied by width, the area of the wall is 8 x 20 = 160 square feet. Multiplying 0.1 x 160 yields $16 as the cost of the paint.

The *average* or *mean* of a collection of numbers is given by adding those numbers together and then dividing by the total number of values. A *weighted average* or *weighted mean* is given by adding the numbers multiplied by their weights, then dividing by the sum of the weights:

$$\frac{w_1 x_1 + w_2 x_2 + w_3 x_3 \dots + w_n x_n}{w_1 + w_2 + w_3 + \dots + w_n}$$

An *ordinary average* is a weighted average where all the weights are 1.

Equations and Inequalities

An equation contains two sides separated by an equal sign. Equations can be solved to determine the value(s) of the variable for which the statement is true.

Assume the sum of a number and 5 is equal to -8 times the number. To find this unknown number, a simple equation can be written to represent the problem. Key words such as difference, equal, and times are used to form the following equation with one variable: $n + 5 = -8n$. When solving for n, opposite operations are used. First, n is subtracted from $-8n$ across the equals sign, resulting in $5 =$

$-9n$. Then, -9 is divided on both sides, leaving $n = -\frac{5}{9}$. This solution can be graphed on the number line with a dot as shown below:

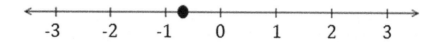

If the problem were changed to say, "The sum of a number and 5 is greater than -8 times the number," then an inequality would be used instead of an equation. Using key words again, *greater than* is represented by the symbol >. The inequality $n + 5 > -8n$ can be solved using the same techniques, resulting in $n < -\frac{5}{9}$. The only time solving an inequality differs from solving an equation is when a negative number is either multiplied times or divided by each side of the inequality. The sign must be switched in this case. For this example, the graph of the solution changes to the following graph because the solution represents all real numbers less than $-\frac{5}{9}$. Not included in this solution is $-\frac{5}{9}$ because it is a *less than* symbol, not *equal to*.

Equations and inequalities in two variables represent a relationship. Jim owns a car wash and charges $40 per car. The rent for the facility is $350 per month. An equation can be written to relate the number of cars Jim cleans to the money he makes per month. Let x represent the number of cars and y represent the profit Jim makes each month from the car wash. The equation $y = 40x - 350$ can be used to show Jim's profit or loss. Since this equation has two variables, the coordinate plane can be used to show the relationship and predict profit or loss for Jim. The following graph shows that Jim must wash at least nine cars to pay the rent, where $x = 9$.

Anything nine cars and above will yield a profit as shown by the value on the y-axis.

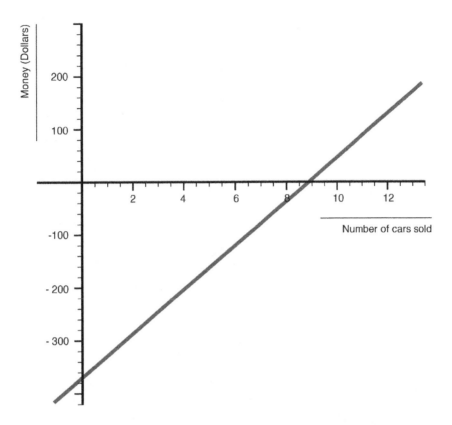

With a single equation in two variables, the solutions are limited only by the situation the equation represents. When two equations or inequalities are used, more constraints are added. For example, in a system of linear equations, there is often—although not always—only one answer. The point of intersection of two lines is the solution. For a system of inequalities, there are infinitely many answers.

The intersection of two solution sets gives the solution set of the system of inequalities. In the following graph, the darker shaded region with the swirls where the shading for the two inequalities overlap. Any set of x and y found in that region satisfies both inequalities. The line with the positive slope is solid, meaning the values on that line are included in the solution.

The line with the negative slope is dotted, so the coordinates on that line are not included.

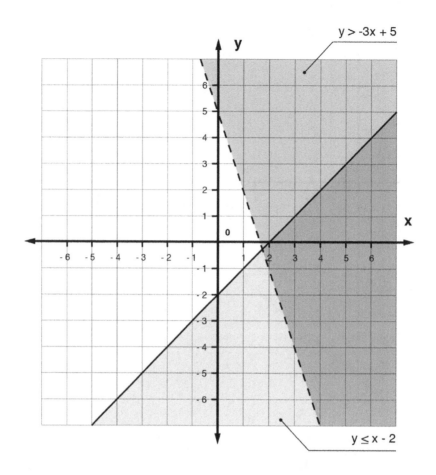

$$y > -3x + 5$$

$$y \leq x - 2$$

Formulas with two variables are equations used to represent a specific relationship. For example, the formula $d = rt$ represents the relationship between distance, rate, and time. If Bob travels at a rate of 35 miles per hour on his road trip from Westminster to Seneca, the formula $d = 35t$ can be used to represent his distance traveled in a specific length of time. Formulas can also be used to show different roles of the variables, transformed without any given numbers. Solving for r, the formula becomes $\frac{d}{t} = r$. The t is moved over by division so that *rate* is a function of distance and time.

Writing Linear Expressions and Equations

Linear expressions and equations are concise mathematical statements that can be written to model a variety of scenarios. Questions found pertaining to this topic will contain one variable only. A variable is an unknown quantity, usually denoted by a letter (x, n, p, etc.). In the case of linear expressions and equations, the power of the variable (its exponent) is 1. A variable without a visible exponent is raised to the first power.

A linear expression is a statement about an unknown quantity expressed in mathematical symbols. The statement "five times a number added to forty" can be expressed as $5x + 40$. A linear equation is a statement in which two expressions (at least one containing a variable) are equal to each other. The

statement "five times a number added to forty is equal to ten" can be expressed as $5x + 40 = 10$. Real-world scenarios can also be expressed mathematically. Consider the following:

Bob had $20 and Tom had $4. After selling 4 ice cream cones to Bob, Tom has as much money as Bob.

The cost of an ice cream cone is an unknown quantity and can be represented by a variable. The amount of money Bob has after his purchase is four times the cost of an ice cream cone subtracted from his original $20. The amount of money Tom has after his sale is four times the cost of an ice cream cone added to his original $4. This can be expressed as: $20 - 4x = 4x + 4$, where x represents the cost of an ice cream cone.

When expressing a verbal or written statement mathematically, it is key to understand words or phrases that can be represented with symbols. The following are examples:

Symbol	Phrase
$+$	added to, increased by, sum of, more than
$-$	decreased by, difference between, less than, take away
x	multiplied by, 3 (4, 5 . . .) times as large, product of
$\div$	divided by, quotient of, half (third, etc.) of
$=$	is, the same as, results in, as much as
$x, t, n, etc.$	a number, unknown quantity, value of

Evaluating and Simplifying Algebraic Expressions

Given an algebraic expression, questions may ask you to evaluate for given values of variable(s). In doing so, you will arrive at a numerical value as an answer. For example:

$$\text{Evaluate } a - 2b + ab \ for \ a = 3 \ and \ b = -1$$

To evaluate an expression, the given values should be substituted for the variables and simplified using the order of operations. In this case: $(3) - 2(-1) + (3)(-1)$. Parentheses are used when substituting.

Given an algebraic expression, questions may ask you to simplify the expression. For example:

$$\text{Simplify } 5x^2 - 10x + 2 - 8x^2 + x - 1.$$

Simplifying algebraic expressions requires combining like terms. A term is a number, variable, or product of a number and variables separated by addition and subtraction. The terms in the above expressions are: $5x^2, -10x, 2, -8x^2, x$, and -1. Like terms have the same variables raised to the same powers (exponents). To combine like terms, the coefficients (numerical factor of the term including sign) are added, while the variables and their powers are kept the same. The example above simplifies to

$$-3x^2 - 9x + 1$$

Solving Equations in One Variable

Solving equations in one variable is the process of isolating a variable on one side of the equation. The letters in an equation and any numbers attached to them are the variables as they stand for unknown quantities that you are trying to solve for. X is commonly used as a variable, though any letter can be used. For example, in $3x - 7 = 20$, the variable is $3x$, and it needs to be isolated. The numbers (also called constants) are -7 and 20. That means $3x$ needs to be on one side of the equals sign (either side is fine), and all the numbers need to be on the other side of the equals sign.

To accomplish this, the equation must be manipulated by performing opposite operations of what already exists. Remember that addition and subtraction are opposites and that multiplication and division are opposites. Any action taken to one side of the equation must be taken on the other side to maintain equality.

So, since the 7 is being subtracted, it can be moved to the right side of the equation by adding seven to both sides:

$$3x - 7 = 20$$

$$3x - 7 + 7 = 20 + 7$$

$$3x = 27$$

Now that the variable $3x$ is on one side and the constants (now combined into one constant) are on the other side, the 3 needs to be moved to the right side. 3 and x are being multiplied together, so 3 then needs to be divided from each side.

$$\frac{3x}{3} = \frac{27}{3}$$

$$x = 9$$

Now that x has been completely isolated, we know its value.

The solution is found to be $x = 9$. This solution can be checked for accuracy by plugging $x = 9$ in the original equation. After simplifying the equation, $20 = 20$ is found, which is a true statement:

$$3 \times 9 - 7 = 20$$

$$27 - 7 = 20$$

$$20 = 20$$

Equations that require solving for a variable (*algebraic equations*) come in many forms. Here are some more examples:

No number attached to the variable:

$$x + 8 = 20$$

$$x + 8 - 8 = 20 - 8$$

$$x = 12$$

Fraction in the variable:

$$\frac{1}{2}z + 24 = 36$$

$$\frac{1}{2}z + 24 - 24 = 36 - 24$$

$$\frac{1}{2}z = 12$$

Now we multiply the fraction by its inverse:

$$\frac{2}{1} \times \frac{1}{2}z = 12 \times \frac{2}{1}$$

$$z = 24$$

Multiple instances of x:

$$14x + x - 4 = 3x + 2$$

All instances of x can be combined.

$$15x - 4 = 3x + 2$$

$$15x - 4 + 4 = 3x + 2 + 4$$

$$15x = 3x + 6$$

$$15x - 3x = 3x + 6 - 3x$$

$$12x = 6$$

$$\frac{12x}{12} = \frac{6}{12}$$

$$x = \frac{1}{2}$$

Writing Linear Inequalities

Linear inequalities and linear equations are both comparisons of two algebraic expressions. However, unlike equations in which the expressions are equal to each other, linear inequalities compare expressions that are unequal. Linear equations typically have one value for the variable that makes the statement true. Linear inequalities generally have an infinite number of values that make the statement true.

Linear inequalities are a concise mathematical way to express the relationship between unequal values. More specifically, they describe in what way the values are unequal. A value could be greater than ($>$); less than ($<$); greater than or equal to ($\geq$); or less than or equal to ($\leq$) another value. The statement "five times a number added to forty is more than sixty-five" can be expressed as $5x + 40 > 65$.

Common words and phrases that express inequalities are:

Symbol	Phrase
<	is under, is below, smaller than, beneath
>	is above, is over, bigger than, exceeds
≤	no more than, at most, maximum
≥	no less than, at least, minimum

Solving Linear Inequalities

When solving a linear inequality, the solution is the set of all numbers that makes the statement true. The inequality $x + 2 \geq 6$ has a solution set of 4 and every number greater than 4 (4.0001, 5, 12, 107, etc.). Adding 2 to 4 or any number greater than 4 would result in a value that is greater than or equal to 6. Therefore, $x \geq 4$ would be the solution set.

Solution sets for linear inequalities often will be displayed using a number line. If a value is included in the set (≥ or ≤), there is a shaded dot placed on that value and an arrow extending in the direction of the solutions. For a variable > or ≥ a number, the arrow would point right on the number line (the direction where the numbers increase); and if a variable is < or ≤ a number, the arrow would point left (where the numbers decrease). If the value is not included in the set (> or <), an open circle on that value would be used with an arrow in the appropriate direction.

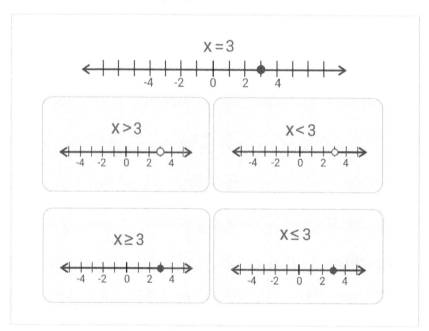

When asked to write a linear inequality given a graph of its solution set, identify whether the value is included (shaded dot or open circle) and the direction in which the arrow is pointing.

In order to algebraically solve a linear inequality, the same steps should be followed as in solving a linear equation. The inequality symbol stays the same for all operations EXCEPT when dividing by a negative number. If dividing by a negative number while solving an inequality, the relationship reverses (the sign

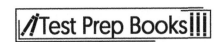

flips). Dividing by a positive does not change the relationship, so the sign stays the same. In other words, > switches to < and vice versa. An example is shown below:

Solve $-2(x + 4) \leq 22$

Distribute: $-2x - 8 \leq 22$

Add 8 to both sides: $-2x \leq 30$

Divide both sides by -2: $x \geq 15$

Functions

A *function* is defined as a relationship between inputs and outputs where there is only one output value for a given input. The input is called the *independent variable*. If the variable is set equal to the output, as in $y = f(x)$, then this is called the *dependent variable*. To indicate the dependent value a function, y, gives for a specific independent variable, x, the notation y = $f(x)$ is used.

As an example, the following function is in function notation:

$$f(x) = 3x - 4$$

The $f(x)$ represents the output value for an input of x. If $x = 2$, the equation becomes:

$$f(2) = 3(2) - 4 = 6 - 4 = 2$$

The input of 2 yields an output of 2, forming the ordered pair $(2, 2)$. The following set of ordered pairs corresponds to the given function: $(2, 2), (0, -4), (-2, -10)$. The set of all possible inputs of a function is its *domain*, and all possible outputs is called the *range*. By definition, each member of the domain is paired with only one member of the range.

Functions can also be defined recursively. In this form, they are not defined explicitly in terms of variables. Instead, they are defined using previously-evaluated function outputs, starting with either $f(0)$ or $f(1)$. An example of a recursively-defined function is:

$$f(1) = 2, f(n) = 2f(n - 1) + 2n, n > 1$$

The domain of this function is the set of all integers.

The domain and range of a function can be found visually by its plot on the coordinate plane. In the function $f(x) = x^2 - 3$, for example, the domain is all real numbers because the parabola stretches as far left and as far right as it can go, with no restrictions. This means that any input value from the real number system will yield an answer in the real number system. For the range, the inequality $y \geq -3$ would be used to describe the possible output values because the parabola has a minimum at $y = -3$. This means there will not be any real output values less than -3 because -3 is the lowest value it reaches on the y-axis.

These same answers for domain and range can be found by observing a table. The table below shows that from input values $x = -1$ to $x = 1$, the output results in a minimum of -3. On each side of $x = 0$, the numbers increase, showing that the range is all real numbers greater than or equal to -3.

x (domain/input)	y (range/output)
-2	1
-1	-2
0	-3
-1	-2
2	1

Different types of functions behave in different ways. The quadratic function described above can be described as decreasing from left to right until the input value of zero, then increasing after that. A function is defined to be increasing over a subset of its domain if for all $x_1 \geq x_2$ in that interval, $f(x_1) \geq f(x_2)$. Also, a function is decreasing over an interval if for all $x_1 \geq x_2$ in that interval:

$$f(x_1) \leq f(x_2)$$

A point in which a function changes from increasing to decreasing can also be labeled as the *maximum value* of a function if it is the largest point the graph reaches on the y-axis. A point in which a function changes from decreasing to increasing can be labeled as the minimum value of a function if it is the smallest point the graph reaches on the y-axis. Maximum values are also known as *extreme values*. The graph of a continuous function does not have any breaks or jumps in the graph. This description is not true of all functions. A radical function, for example, $f(x) = \sqrt{x}$, has a restriction for the domain and range because there are no real negative inputs or outputs for this function. The domain can be stated as $x \geq 0$, and the range is $y \geq 0$.

A piecewise-defined function also has a different appearance on the graph. In the following function, there are three equations defined over different intervals. It is a function because there is only one y-value for each x-value, passing the Vertical Line Test. The domain is all real numbers less than or equal to 6. The range is all real numbers greater than 0. From left to right, the graph decreases to 0, then increases to almost 4, and then jumps to 6.

From input values greater than 2, the input decreases just below 8 to 4, and then stops.

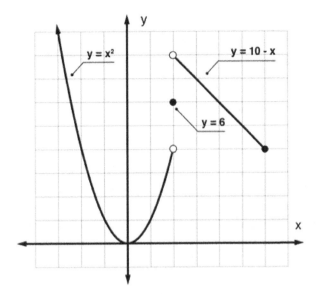

Logarithmic and exponential functions also have different behavior than other functions. These two types of functions are inverses of each other. The *inverse* of a function can be found by switching the place of x and y, and solving for y. When this is done for the exponential equation, $y = 2^x$, the function $y = \log_2 x$ is found. The general form of a *logarithmic function* is $y = \log_b x$, which says b raised to the y power equals x.

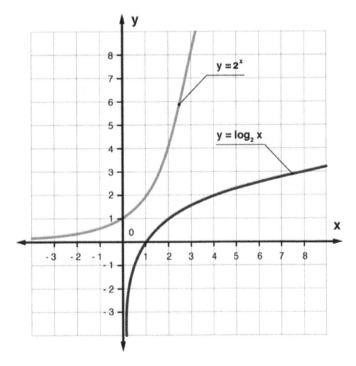

The thick black line on the graph above represents the logarithmic function $y = \log_2 x$. This curve passes through the point $(1, 0)$, just as all log functions do, because any value $b^0 = 1$. The graph of this logarithmic function starts very close to zero, but does not touch the y-axis. The output value will never be zero by the definition of logarithms. The thinner gray line seen above represents the exponential function $y = 2^x$. The behavior of this function is opposite the logarithmic function because the graph of an inverse function is the graph of the original function flipped over the line $y = x$. The curve passes through the point $(0, 1)$ because any number raised to the zero power is one. This curve also gets very close to the x-axis but never touches it because an exponential expression never has an output of zero. The x-axis on this graph is called a horizontal asymptote. An *asymptote* is a line that represents a boundary for a function. It shows a value that the function will get close to, but never reach.

Functions can also be described as being even, odd, or neither. If $f(-x) = f(x)$, the function is even. For example, the function:

$$f(x) = x^2 - 2$$

is even. Plugging in $x = 2$ yields an output of $y = 2$. After changing the input to $x = -2$, the output is still $y = 2$. The output is the same for opposite inputs. Another way to observe an even function is by the symmetry of the graph. If the graph is symmetrical about the axis, then the function is even. If the graph is symmetric about the origin, then the function is odd. Algebraically, if

$$f(-x) = -f(x)$$

the function is odd.

Finally, a function can be described as *periodic* if it repeats itself in regular intervals. Common periodic functions are trigonometric functions. For example, $y = \sin x$ is a periodic function with period 2π because it repeats itself every 2π units along the x-axis.

Building Functions

Functions can be built out of the context of a situation. For example, the relationship between the money paid for a gym membership and the months that someone has been a member can be described through a function. If the one-time membership fee is $40 and the monthly fee is $30, then the function can be written $f(x) = 30x + 40$. The x-value represents the number of months the person has been part of the gym, while the output is the total money paid for the membership. The table below shows this relationship. It is a representation of the function because the initial cost is $40 and the cost increases each month by $30.

x (months)	y (money paid to gym)
0	40
1	70
2	100
3	130

Functions can also be built from existing functions. For example, a given function $f(x)$ can be transformed by adding a constant, multiplying by a constant, or changing the input value by a constant. The new function $g(x) = f(x) + k$ represents a vertical shift of the original function. In $f(x) = 3x - 2$, a vertical shift 4 units up would be:

$$g(x) = 3x - 2 + 4 = 3x + 2$$

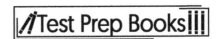

Multiplying the function, times a constant k represents a vertical stretch, based on whether the constant is greater than or less than 1. The following function represents a stretch:

$$g(x) = kf(x) = 4(3x - 2) = 12x - 8$$

Changing the input x, by a constant, forms the function:

$$g(x) = f(x + k) = 3(x + 4) - 2$$

$$3x + 12 - 2 = 3x + 10$$

and this represents a horizontal shift to the left 4 units. If $(x - 4)$ was plugged into the function, it would represent a vertical shift.

Graphing Linear Functions

A function is called *linear* if it can take the form of the equation $f(x) = ax + b$, or $y = ax + b$, for any two numbers a and b. A linear equation forms a straight line when graphed on the coordinate plane. An example of a linear function is shown below on the graph.

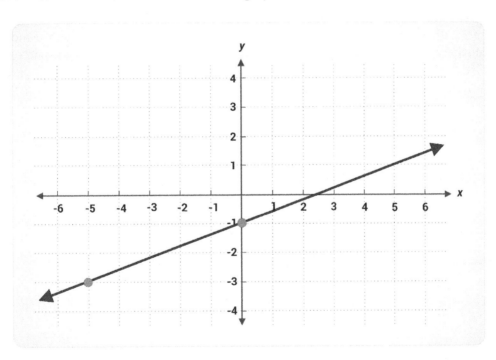

This is a graph of the following function: $y = \frac{2}{5}x - 1$. A table of values that satisfies this function is shown below.

x	y
-5	-3
0	-1
5	1
10	3

These points can be found on the graph using the form (x,y).

To graph relations and functions, the Cartesian plane is used. This means to think of the plane as being given a grid of squares, with one direction being the x-axis and the other direction the y-axis. Generally, the independent variable is placed along the horizontal axis, and the dependent variable is placed along the vertical axis. Any point on the plane can be specified by saying how far to go along the x-axis and how far along the y-axis with a pair of numbers (x, y). Specific values for these pairs can be given names such as $C = (-1, 3)$. Negative values mean to move left or down; positive values mean to move right or up. The point where the axes cross one another is called the *origin*. The origin has coordinates $(0, 0)$ and is usually called O when given a specific label. An illustration of the Cartesian plane, along with graphs of $(2, 1)$ and $(-1, -1)$, are below.

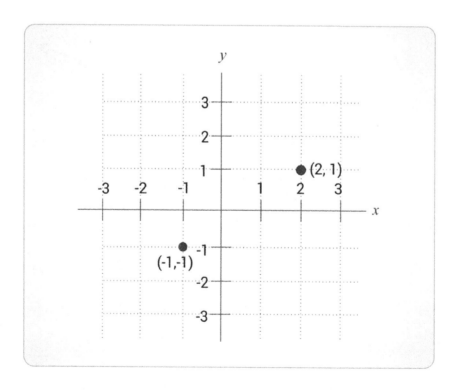

Relations also can be graphed by marking each point whose coordinates satisfy the relation. If the relation is a function, then there is only one value of y for any given value of x. This leads to the **vertical line test**: if a relation is graphed, then the relation is a function if any possible vertical line drawn

anywhere along the graph would only touch the graph of the relation in no more than one place. Conversely, when graphing a function, then any possible vertical line drawn will not touch the graph of the function at any point or will touch the function at just one point. This test is made from the definition of a function, where each x-value must be mapped to one and only one y-value.

When graphing a linear function, note that the ratio of the change of the y coordinate to the change in the x coordinate is constant between any two points on the resulting line, no matter which two points are chosen. In other words, in a pair of points on a line, (x_1, y_1) and (x_2, y_2), with $x_1 \neq x_2$ so that the two points are distinct, then the ratio $\frac{y_2 - y_1}{x_2 - x_1}$ will be the same, regardless of which particular pair of points are chosen. This ratio, $\frac{y_2 - y_1}{x_2 - x_1}$, is called the *slope* of the line and is frequently denoted with the letter m. If slope m is positive, then the line goes upward when moving to the right, while if slope m is negative, then the line goes downward when moving to the right. If the slope is 0, then the line is called *horizontal*, and the y coordinate is constant along the entire line. In lines where the x coordinate is constant along the entire line, y is not actually a function of x. For such lines, the slope is not defined. These lines are called *vertical* lines.

Linear functions may take forms other than $y = ax + b$. The most common forms of linear equations are explained below:

1. Standard Form: $Ax + By = C$, in which the slope is given by $m = \frac{-A}{B}$, and the y-intercept is given by $\frac{C}{B}$.

2. Slope-Intercept Form: $y = mx + b$, where the slope is m and the y intercept is b.

3. Point-Slope Form: $y - y_1 = m(x - x_1)$, where the slope is m and (x_1, y_1) is any point on the chosen line.

4. Two-Point Form: $\frac{y - y_1}{x - x_1} = \frac{y_2 - y_1}{x_2 - x_1}$, where (x_1, y_1) and (x_2, y_2) are any two distinct points on the chosen line. Note that the slope is given by $m = \frac{y_2 - y_1}{x_2 - x_1}$.

5. Intercept Form: $\frac{x}{x_1} + \frac{y}{y_1} = 1$, in which x_1 is the x-intercept and y_1 is the y-intercept.

These five ways to write linear equations are all useful in different circumstances. Depending on the given information, it may be easier to write one of the forms over another.

If $y = mx$, y is directly proportional to x. In this case, changing x by a factor changes y by that same factor. If $y = \frac{m}{x}$, y is inversely proportional to x. For example, if x is increased by a factor of 3, then y will be decreased by the same factor, 3.

The *midpoint* between two points, (x_1, y_1) and (x_2, y_2), is given by taking the average of the x coordinates and the average of the y coordinates:

$$\left(\frac{x_1 + x_2}{2}, \frac{y_1 + y_2}{2} \right)$$

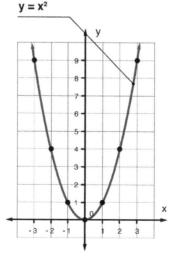

The *distance* between two points, (x_1, y_1) and (x_2, y_2), is given by the *Pythagorean formula*:

$$\sqrt{(x_2 - x_1)^2 + (y_2 - y_1)^2}$$

To find the perpendicular distance between a line $Ax + By = C$ and a point (x_1, y_1) not on the line, we need to use the formula

$$\frac{|Ax_1 + By_1 + C|}{\sqrt{A^2 + B^2}}$$

Common Functions

Three common functions used to model different relationships between quantities are linear, quadratic, and exponential functions. Linear functions are the simplest of the three, and the independent variable x has an exponent of 1. Written in the most common form, $y = mx + b$, the coefficient of x tells how fast the function grows at a constant rate, and the b-value tells the starting point. A quadratic function has an exponent of 2 on the independent variable x. Standard form for this type of function is $y = ax^2 + bx + c$, and the graph is a parabola. These type functions grow at a changing rate. An exponential function has an independent variable in the exponent $y = ab^x$. The graph of these types of functions is described as *growth* or *decay*, based on whether the base, b, is greater than or less than 1. These functions are different from quadratic functions because the base stays constant. A common base is base e.

The following three functions model a linear, quadratic, and exponential function respectively: $y = 2x$, $y = x^2$, and $y = 2^x$. Their graphs are shown below. The first graph, modeling the linear function, shows that the growth is constant over each interval. With a horizontal change of 1, the vertical change is 2. It models a constant positive growth. The second graph shows the quadratic function, which is a curve that is symmetric across the y-axis. The growth is not constant, but the change is mirrored over the axis. The last graph models the exponential function, where the horizontal change of 1 yields a vertical change that increases more and more. The exponential graph gets very close to the x-axis, but never touches it, meaning there is an asymptote there. The y-value can never be zero because the base of 2 can never be raised to an input value that yields an output of zero.

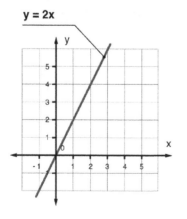

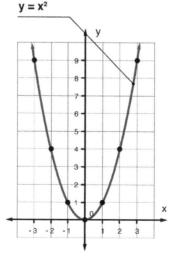

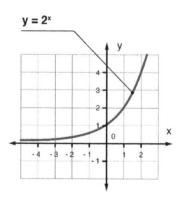

The three tables below show specific values for three types of functions. The third column in each table shows the change in the y-values for each interval. The first table shows a constant change of 2 for each equal interval, which matches the slope in the equation $y = 2x$. The second table shows an increasing change, but it also has a pattern. The increase is changing by 2 more each time, so the change is quadratic. The third table shows the change as factors of the base, 2. It shows a continuing pattern of factors of the base.

y = 2x		
x	y	$\triangle y$
1	2	
2	4	2
3	6	2
4	8	2
5	10	2

y = x²		
x	y	$\triangle y$
1	1	
2	4	3
3	9	5
4	16	7
5	25	9

y = 2ˣ		
x	y	$\triangle y$
1	2	
2	4	2
3	8	4
4	16	8
5	32	16

Given a table of values, the type of function can be determined by observing the change in y over equal intervals. For example, the tables below model two functions. The changes in interval for the x-values is 1 for both tables. For the first table, the y-values increase by 5 for each interval. Since the change is constant, the situation can be described as a linear function. The equation would be $y = 5x + 3$. For the second table, the change for y is 5, 20, 100, and 500, respectively. The increases are multiples of 5, meaning the situation can be modeled by an exponential function. The equation $y = 5^x + 3$ models this situation.

x	y
0	3
1	8
2	13
3	18
4	23

x	y
0	3
1	8
2	28
3	128
4	628

Quadratic equations can be used to model real-world area problems. For example, a farmer may have a rectangular field that he needs to sow with seed. The field has length $x + 8$ and width $2x$. The formula for area should be used: $A = lw$. Therefore:

$$A = (x + 8) * 2x = 2x^2 + 16x$$

The possible values for the length and width can be shown in a table, with input x and output A. If the equation was graphed, the possible area values can be seen on the y-axis for given x-values.

Exponential growth and decay can be found in real-world situations. For example, if a piece of notebook paper is folded 25 times, the thickness of the paper can be found. To model this situation, a table can be used. The initial point is one-fold, which yields a thickness of 2 papers. For the second fold, the thickness is 4. Since the thickness doubles each time, the table below shows the thickness for the next few folds. Notice the thickness changes by the same factor each time. Since this change for a constant interval of folds is a factor of 2, the function is exponential. The equation for this is $y = 2^x$. For twenty-five folds, the thickness would be 33,554,432 papers.

x (folds)	y (paper thickness)
0	1
1	2
2	4
3	8
4	16
5	32

One exponential formula that is commonly used is the *interest formula*: $A = Pe^{rt}$. In this formula, interest is compounded continuously. A is the value of the investment after the time, t, in years. P is the initial amount of the investment, r is the interest rate, and e is the constant equal to approximately 2.718. Given an initial amount of $200 and a time of 3 years, if interest is compounded continuously at a rate of 6%, the total investment value can be found by plugging each value into the formula. The invested value at the end is $239.44. In more complex problems, the final investment may be given, and the rate may be the unknown. In this case, the formula becomes:

$$239.44 = 200e^{r3}$$

Solving for r requires isolating the exponential expression on one side by dividing by 200, yielding the equation $1.20 = e^{r3}$. Taking the natural log of both sides results in $\ln(1.2) = r3$. Using a calculator to evaluate the logarithmic expression, $r = 0.06 = 6\%$.

Patterns

Patterns within a sequence can come in 2 distinct forms: the items (shapes, numbers, etc.) either repeat in a constant order, or the items change from one step to another in some consistent way. The core is the smallest unit, or number of items, that repeats in a repeating pattern. For example, the pattern oo▲oo▲o... has a core that is oo▲. Knowing only the core, the pattern can be extended. Knowing the number of steps in the core allows the identification of an item in each step without drawing/writing the entire pattern out. For example, suppose the tenth item in the previous pattern must be determined. Because the core consists of three items (oo▲), the core repeats in multiples of 3. In other words, steps 3, 6, 9, 12, etc. will be ▲ completing the core with the core starting over on the next step. For the above example, the 9th step will be ▲ and the 10th will be o.

The most common patterns in which each item changes from one step to the next are arithmetic and geometric sequences. An arithmetic sequence is one in which the items increase or decrease by a constant difference. In other words, the same thing is added or subtracted to each item or step to produce the next. To determine if a sequence is arithmetic, determine what must be added or subtracted to step one to produce step two. Then, check if the same thing is added/subtracted to step two to produce step three. The same thing must be added/subtracted to step three to produce step four, and so on. Consider the pattern 13, 10, 7, 4 . . . To get from step one (13) to step two (10) by adding or subtracting requires subtracting by 3. The next step is checking if subtracting 3 from step two (10) will produce step three (7), and subtracting 3 from step three (7) will produce step four (4). In this

case, the pattern holds true. Therefore, this is an arithmetic sequence in which each step is produced by subtracting 3 from the previous step. To extend the sequence, 3 is subtracted from the last step to produce the next. The next three numbers in the sequence are 1, -2, -5.

A geometric sequence is one in which each step is produced by multiplying or dividing the previous step by the same number. To determine if a sequence is geometric, decide what step one must be multiplied or divided by to produce step two. Then check if multiplying or dividing step two by the same number produces step three, and so on. Consider the pattern 2, 8, 32, 128 . . . To get from step one (2) to step two (8) requires multiplication by 4. The next step determines if multiplying step two (8) by 4 produces step three (32), and multiplying step three (32) by 4 produces step four (128). In this case, the pattern holds true. Therefore, this is a geometric sequence in which each step is produced by multiplying the previous step by 4. To extend the sequence, the last step is multiplied by 4 and repeated. The next three numbers in the sequence are 512; 2,048; 8,192.

Although arithmetic and geometric sequences typically use numbers, these sequences can also be represented by shapes. For example, an arithmetic sequence could consist of shapes with three sides, four sides, and five sides (add one side to the previous step to produce the next). A geometric sequence could consist of eight blocks, four blocks, and two blocks (each step is produced by dividing the number of blocks in the previous step by 2).

Data Analysis

Interpretation of Graphs

Data can be represented in many ways including picture graphs, bar graphs, line plots, and tally charts. It is important to be able to organize the data into categories that could be represented using one of these methods. Equally important is the ability to read these types of diagrams and interpret their meaning.

A *picture graph* is a diagram that shows pictorial representation of data being discussed. The symbols used can represent a certain number of objects.

Notice how each fruit symbol in the following graph represents a count of two fruits. One drawback of picture graphs is that they can be less accurate if each symbol represents a large number. For example,

if each banana symbol represented ten bananas, and students consumed 22 bananas, it may be challenging to draw and interpret two and one-fifth bananas as a frequency count of 22.

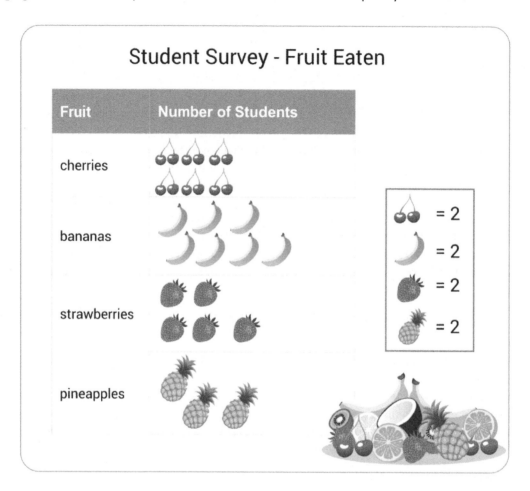

A *bar graph* is a diagram in which the quantity of items within a specific classification is represented by the height of a rectangle. Each type of classification is represented by a rectangle of equal width. Here is an example of a bar graph:

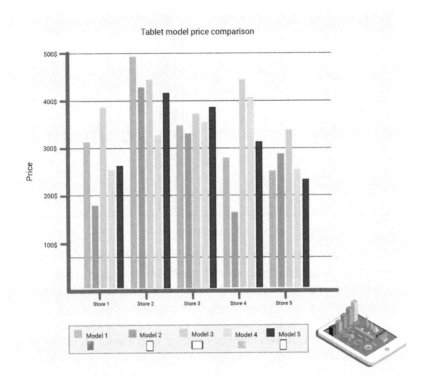

A *line plot* is a diagram that shows quantity of data along a number line. It is a quick way to record data in a structure similar to a bar graph without needing to do the required shading of a bar graph. Here is an example of a line plot:

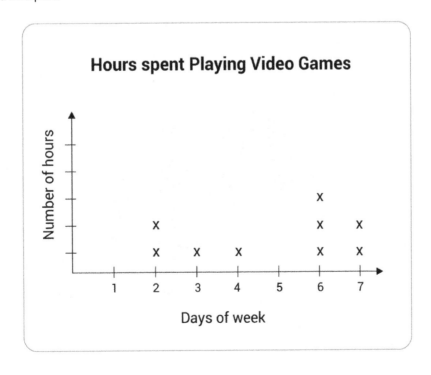

A *tally chart* is a diagram in which tally marks are utilized to represent data. Tally marks are a means of showing a quantity of objects within a specific classification. Here is an example of a tally chart:

Number of days with rain	Number of weeks
0	II
1	HHT
2	HHT
3	HHT
4	HHT HHT HHT IIII
5	HHT I
6	HHT I
7	IIII

Data is often recorded using fractions, such as half a mile, and understanding fractions is critical because of their popular use in real-world applications. Also, it is extremely important to label values with their units when using data. For example, regarding length, the number 2 is meaningless unless it is attached to a unit. Writing 2 cm shows that the number refers to the length of an object.

A circle graph, also called a pie chart, shows categorical data with each category representing a percentage of the whole data set. To make a circle graph, the percent of the data set for each category must be determined. To do so, the frequency of the category is divided by the total number of data points and converted to a percent. For example, if 80 people were asked what their favorite sport is and 20 responded basketball, basketball makes up 25% of the data ($\frac{20}{80} = .25 = 25\%$). Each category in a data set is represented by a *slice* of the circle proportionate to its percentage of the whole.

FAVORITE SPORT

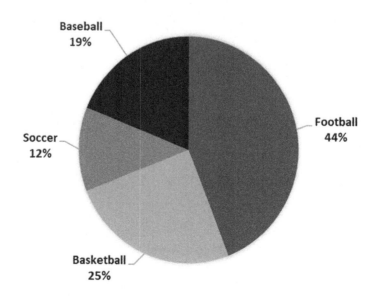

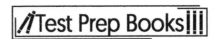

A scatter plot displays the relationship between two variables. Values for the independent variable, typically denoted by *x*, are paired with values for the dependent variable, typically denoted by *y*. Each set of corresponding values are written as an ordered pair (*x*, *y*). To construct the graph, a coordinate grid is labeled with the *x*-axis representing the independent variable and the *y*-axis representing the dependent variable. Each ordered pair is graphed.

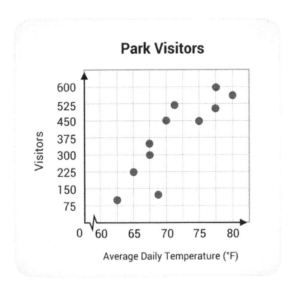

Like a scatter plot, a line graph compares two variables that change continuously, typically over time. Paired data values (ordered pair) are plotted on a coordinate grid with the *x*- and *y*-axis representing the two variables. A line is drawn from each point to the next, going from left to right. A double line graph simply displays two sets of data that contain values for the same two variables. The double line graph below displays the profit for given years (two variables) for Company A and Company B (two data sets).

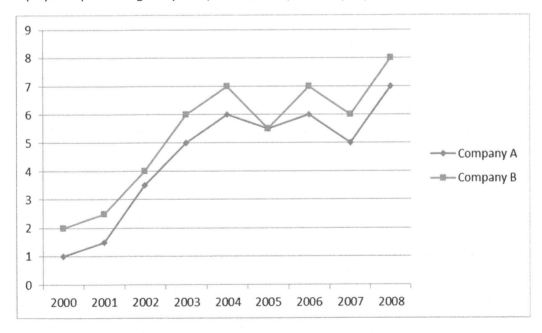

Choosing the appropriate graph to display a data set depends on what type of data is included in the set and what information must be shown. Histograms and box plots can be used for data sets consisting of individual values across a wide range. Examples include test scores and incomes. Histograms and box plots will indicate the center, spread, range, and outliers of a data set. A histogram will show the shape of the data set, while a box plot will divide the set into quartiles (25% increments), allowing for comparison between a given value and the entire set.

Scatter plots and line graphs can be used to display data consisting of two variables. Examples include height and weight, or distance and time. A correlation between the variables is determined by examining the points on the graph. Line graphs are used if each value for one variable pairs with a distinct value for the other variable. Line graphs show relationships between variables.

Explaining the Relationship between Two Variables

In an experiment, variables are the key to analyzing data, especially when data is in a graph or table. Variables can represent anything, including objects, conditions, events, and amounts of time.

Constant variables remain unchanged by the scientist across all trials. Because they are held constant for all groups in an experiment, they aren't being measured in the experiment, and they are usually ignored. Constants can either be controlled by the scientist directly like the nutrition, water, and sunlight given to plants, or they can be selected by the scientist specifically for an experiment like using a certain animal species or choosing to investigate only people of a certain age group.

Independent variables are also controlled by the scientist, but they are the same only for each group or trial in the experiment. Each group might be composed of students that all have the same color of car or each trial may be run on different soda brands. The independent variable of an experiment is what is being indirectly tested because it causes change in the dependent variables.

Dependent variables experience change caused by the independent variable and are what is being measured or observed. For example, college acceptance rates could be a dependent variable of an experiment that sorted a large sample of high school students by an independent variable such as test scores. In this experiment, the scientist groups the high school students by the independent variable (test scores) to see how it affects the dependent variable (their college acceptance rates).

Note that most variables can be held constant in one experiment but independent or dependent in another. For example, when testing how well a fertilizer aids plant growth, its amount of sunlight should be held constant for each group of plants, but if the experiment is being done to determine the proper amount of sunlight a plant should have, the amount of sunlight is an independent variable because it is necessarily changed for each group of plants.

Probability and Statistics

Probability

There are many events that cannot be predicted with guaranteed certainty or that do not have a definite outcome. Instead, only the likeliness of a particular outcome can be expressed. This concept—how likely something is to happen—deals with the concept of *probability*. Consider what happens when a coin is tossed. It can either land with heads facing up or tails, which means that there are two possible outcomes of the toss. Mathematicians would say that the probability of getting heads is $\frac{1}{2}$ because there are two possible outcomes, heads or tails, and heads is one of those two. Therefore, it is expected that heads will land face up one time (the numerator) for every two tosses (the denominator). The same can

be said for tails, so the probability is also $\frac{1}{2}$ for getting tails. A similar concept exists with rolling a single die. There are six faces on the die, one with each of the following numbers: 1, 2, 3, 4, 5, 6. Therefore, with any roll, the probability of rolling any of these numbers, for example, a 3, is $\frac{1}{6}$ because there is one side with a 3 and six total sides.

The examples of the coin toss and rolling a die model the general formula for determining the probability of a chance event happening:

$$\text{Probability of an event occurring} = \frac{\#\ of\ possible\ ways\ it\ can\ happen}{total\ number\ of\ possible\ outcomes}$$

Consider a bag of lollipops. There are three flavors of candy: strawberry, apple, and grape. The bag of lollipops contains 1 strawberry, 2 apple, and 5 grape ones. What is the probability that an apple lollipop will be drawn from the bag at random? The first step is to find the total number of lollipops in the bag, which is the sum of the number of each of the three flavors: 1 strawberry, 2 apple, and 5 grape = 8 lollipops. Then, the probability of randomly drawing an apple pop is calculated as follows:

$$\text{Probability of picking apple} = \frac{2\ apple\ pops}{8\ total\ pops}$$

So, the probability of getting an apple pop is $\frac{2}{8}$, which can be simplified to $\frac{1}{4}$, or 0.25. Notice how the probability is converted to a decimal. All probabilities can be expressed with a number between 0 and 1 (usually a decimal somewhere in this range). A probability of zero means that there is no chance that the event will occur. In the example of the lollipops, the probability of pulling an orange-flavored candy is zero because there are no orange pops in the bag. On the other end of the spectrum, a probability of 1 is equivalent to 100% or a definite outcome. For example, if a single-flavor bag of lollipops is purchased, for example a bag containing 8 cherry lollipops, the probability of randomly selecting a cherry candy is $\frac{8}{8}$, which equals 1 (since a fraction where the numerator and denominator are the same is equivalent to 1).

It is important to note that except in the case of a probability of 0 or 1, the calculated probability is just a guide or expectation as to the likelihood of a certain outcome occurring. Just because a coin has two sides, it doesn't guarantee that one out of every two tosses (a probability of 0.50) will land with heads facing up. Sometimes, two or more tosses in a row will be heads; the same can be said for tails. Out of 100 tosses, the probability of 0.50 means that we expect 50 tosses to land on heads. If we actually carry out the 100 coin flips, we might get 49 heads, 47 heads, 56 heads, etc. However, it is highly likely that the number will be close to 50.

Conditional probability is an important concept because, in many situations, the likelihood of one outcome can differ radically depending on how something else comes out. The probability of passing a test given that one has studied all of the material is generally much higher than the probability of passing a test given that one has not studied at all. The probability of a person having heart trouble is much lower if that person exercises regularly. The probability that a college student will graduate is higher when his or her SAT scores are higher, and so on. For this reason, there are many people who are interested in conditional probabilities.

Note that in some practical situations, changing the order of the conditional probabilities can make the outcome very different. For example, the probability that a person with heart trouble has exercised regularly is quite different than the probability that a person who exercises regularly will have heart trouble. The probability of a person receiving a military-only award, given that he or she is or was a soldier, is generally not very high, but the probability that a person being or having been a soldier, given that he or she received a military-only award, is 1.

However, in some cases, the outcomes do not influence one another this way. If the probability of a given outcome is the same regardless of the previous outcome, then it's considered an *independent* probability. Our examples of the coin toss or rolling a die are both examples of independent probabilities. If the same die is rolled repeatedly, then the next number rolled should not depend on which numbers have been rolled previously. Similarly, if a coin is flipped, then the next flip's outcome does not depend on the outcomes of previous flips.

This can sometimes be counterintuitive, since when rolling a die or flipping a coin, there can be a streak of surprising results. If, however, it is known that the die or coin is fair, then these results are just the result of the fact that over long periods of time, it is very likely that some unlikely streaks of outcomes will occur. Therefore, avoid making the mistake of thinking that when considering a series of independent outcomes, a particular outcome is "due to happen" simply because a surprising series of outcomes has already been seen.

Statistical Questions

A statistical question is answered by collecting data with variability. Data consists of facts and/or statistics (numbers), and variability refers to a tendency to shift or change. Data is a broad term, inclusive of things like height, favorite color, name, salary, temperature, gas mileage, and language. Questions requiring data as an answer are not necessarily statistical questions. If there is no variability in the data, then the question is not statistical in nature. Consider the following examples: what is Mary's favorite color? How much money does your mother make? What was the highest temperature last week? How many miles did your car get on its last tank of gas? How much taller than Bob is Ed?

None of the above are statistical questions because each case lacks variability in the data needed to answer the question. The questions on favorite color, salary, and gas mileage each require a single piece of data, whether a fact or statistic. Therefore, variability is absent. Although the temperature question requires multiple pieces of data (the high temperature for each day), a single, distinct number is the answer. The height question requires two pieces of data, Bob's height and Ed's height, but no difference in variability exists between those two values. Therefore, this is not a statistical question. Statistical questions typically require calculations with data.

Consider the following statistical questions:

How many miles per gallon of gas does the 2016 Honda Civic get? To answer this question, data must be collected. This data should include miles driven and gallons used. Different cars, different drivers, and different driving conditions will produce different results. Therefore, variability exists in the data. To answer the question, the mean (average) value could be determined.

Are American men taller than German men? To answer this question, data must be collected. This data should include the heights of American men and the heights of German men. All American men are not the same height and all German men are not the same height. Some American men are taller than some German men and some German men are taller than some American men. Therefore, variability exists in

the data. To answer the question, the median values for each group could be determined and compared.

The following are more examples of statistical questions: What proportion of 4th graders have a favorite color of blue? How much money do teachers make? Is it colder in Boston or Chicago?

Data Gathering Techniques

Statistics involves making decisions and predictions about larger sets of data based on smaller data sets. The information from a small subset can help predict what happens in the entire set. The smaller data set is called a *sample* and the larger data set for which the decision is being made is called a *population*. The three most common types of data gathering techniques are sample surveys, experiments, and observational studies. *Sample surveys* involve collecting data from a random sample of people from a desired population. The measurement of the variable is only performed on this set of people. To have accurate data, the sampling must be unbiased and random. For example, surveying students in an advanced calculus class on how much they enjoy math classes is not a useful sample if the population should be all college students based on the research question. There are many methods to form a random sample, and all adhere to the fact that every sample that could be chosen has a predetermined probability of being chosen. Once the sample is chosen, statistical experiments can then be carried out to investigate real-world problems.

An *experiment* is the method in which a hypothesis is tested using a trial-and-error process. A cause and the effect of that cause are measured, and the hypothesis is accepted or rejected. Experiments are usually completed in a controlled environment where the results of a control population are compared to the results of a test population. The groups are selected using a randomization process in which each group has a representative mix of the population being tested. Finally, an *observational study* is similar to an experiment. However, this design is used when there cannot be a designed control and test population because of circumstances (e.g., lack of funding or unrealistic expectations). Instead, existing control and test populations must be used, so this method has a lack of randomization.

To make decisions concerning populations, data must be collected from a sample. The sample must be large enough to be able to make conclusions. A common way to collect data is via surveys and polls. Every survey and poll must be designed so that there is no bias. An example of a biased survey is one with loaded questions, which are either intentionally worded or ordered to obtain a desired response. Once the data is obtained, conclusions should not be made that are not justified by statistical analysis. One must make sure the difference between correlation and causation is understood. Correlation implies there is an association between two variables, and correlation does not imply causation.

Measures of Center and Range

The center of a set of data (statistical values) can be represented by its mean, median, or mode. These are sometimes referred to as measures of central tendency. The mean is the average of the data set. The mean can be calculated by adding the data values and dividing by the sample size (the number of data points). Suppose a student has test scores of 93, 84, 88, 72, 91, and 77. To find the mean, or average, the scores are added and the sum is divided by 6 because there are 6 test scores:

$$\frac{93 + 84 + 88 + 72 + 91 + 77}{6} = \frac{505}{6} = 84.17$$

Given the mean of a data set and the sum of the data points, the sample size can be determined by dividing the sum by the mean. Suppose you are told that Kate averaged 12 points per game and scored a total of 156 points for the season. The number of games that she played (the sample size or the number of data points) can be determined by dividing the total points (sum of data points) by her average (mean of data points): $\frac{156}{12} = 13$. Therefore, Kate played in 13 games this season.

If given the mean of a data set and the sample size, the sum of the data points can be determined by multiplying the mean and sample size. Suppose you are told that Tom worked 6 days last week for an average of 5.5 hours per day. The total number of hours worked for the week (sum of data points) can be determined by multiplying his daily average (mean of data points) by the number of days worked (sample size): $5.5 \times 6 = 33$. Therefore, Tom worked a total of 33 hours last week.

The median of a data set is the value of the data point in the middle when the sample is arranged in numerical order. To find the median of a data set, the values are written in order from least to greatest. The lowest and highest values are simultaneously eliminated, repeating until the value in the middle remains. Suppose the salaries of math teachers are: $35,000; $38,500; $41,000; $42,000; $42,000; $44,500; $49,000. The values are listed from least to greatest to find the median. The lowest and highest values are eliminated until only the middle value remains. Repeating this step three times reveals a median salary of $42,000. If the sample set has an even number of data points, two values will remain after all others are eliminated. In this case, the mean of the two middle values is the median. Consider the following data set: 7, 9, 10, 13, 14, 14. Eliminating the lowest and highest values twice leaves two values, 10 and 13, in the middle. The mean of these values $\left(\frac{10+13}{2}\right)$ is the median. Therefore, the set has a median of 11.5.

The mode of a data set is the value that appears most often. A data set may have a single mode, multiple modes, or no mode. If different values repeat equally as often, multiple modes exist. If no value repeats, no mode exists. Consider the following data sets:

> A: 7, 9, 10, 13, 14, 14
> B: 37, 44, 33, 37, 49, 44, 51, 34, 37, 33, 44
> C: 173, 154, 151, 168, 155

Set A has a mode of 14. Set B has modes of 37 and 44. Set C has no mode.

The range of a data set is the difference between the highest and the lowest values in the set. The range can be considered the span of the data set. To determine the range, the smallest value in the set is subtracted from the largest value. The ranges for the data sets A, B, and C above are calculated as follows: A: $14 - 7 = 7$; B: $51 - 33 = 18$; C: $173 - 151 = 22$.

Geometry

Lines and Planes

The basic unit of geometry is a point. These are locations on the plane that have no width or breadth. The position of a point is indicated with a dot and usually named with a single uppercase letter, such as

point *A* or point *T*. A point is a place, not a thing, and therefore has no dimensions or size. A set of points that lies on the same line is called collinear. A set of points that lies on the same plane is called coplanar.

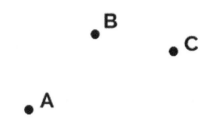

The image above displays point *A*, point *B*, and point *C*.

Any pair of points *A*, *B* on the plane will determine a unique straight line between them. This line is denoted *AB*. Sometimes to emphasize a line is being considered, this will be written as $\overleftrightarrow{AB}$.

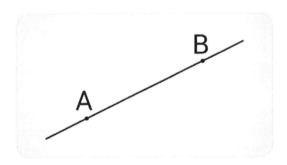

If the Cartesian coordinates for *A* and *B* are known, then the distance $d(A, B)$ along the line between them can be measured using the *Pythagorean formula*, which states that if $A = (x_1, y_1)$ and $B = (x_2, y_2)$, then the distance between them is $d(A, B) = \sqrt{(x_2 - x_1)^2 + (y_2 - y_1)^2}$.

The part of a line that lies between *A* and *B* is called a *line segment*. It has two endpoints, one at *A* and one at *B*. *Rays* also can be formed. Given points *A* and *B*, a *ray* is the portion of a line that starts at one of these points, passes through the other, and keeps on going. Therefore, a ray has a single endpoint, but the other end goes off to infinity.

Two lines are considered parallel to each other if, while extending infinitely, they will never intersect (or meet). Parallel lines point in the same direction and are always the same distance apart. Two lines are considered perpendicular if they intersect to form right angles. Right angles are 90°. Typically, a small box is drawn at the intersection point to indicate the right angle.

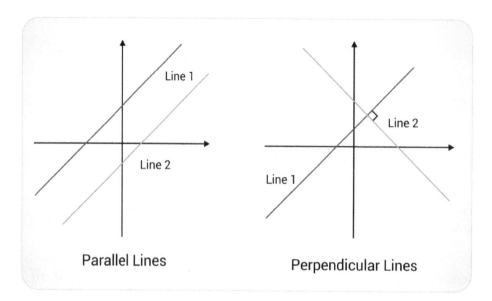

Parallel Lines   Perpendicular Lines

Line 1 is parallel to line 2 in the left image and is written as line 1 || line 2. Line 1 is perpendicular to line 2 in the right image and is written as line 1 ⊥ line 2.

Angles

An angle consists of two rays that have a common endpoint. This common endpoint is called the vertex of the angle. The two rays can be called sides of the angle. The angle below has a vertex at point *B* and the sides consist of ray *BA* and ray *BC*. An angle can be named in three ways:

1. Using the vertex and a point from each side, with the vertex letter in the middle.
2. Using only the vertex. This can only be used if it is the only angle with that vertex.
3. Using a number that is written inside the angle.

The angle below can be written $\angle ABC$ (read angle *ABC*), $\angle CBA$, $\angle B$, or $\angle 1$.

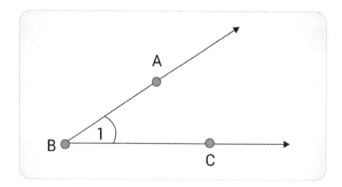

An angle divides a plane, or flat surface, into three parts: the angle itself, the interior (inside) of the angle, and the exterior (outside) of the angle. The figure below shows point *M* on the interior of the angle and point *N* on the exterior of the angle.

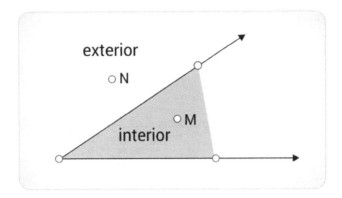

Angles can be measured in units called degrees, with the symbol °. The degree measure of an angle is between 0° and 180° and can be obtained by using a protractor.

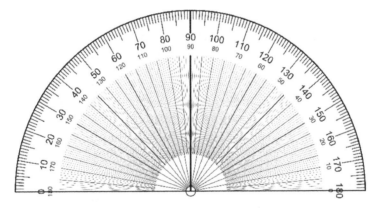

A straight angle (or simply a line) measures exactly 180°. A right angle's sides meet at the vertex to create a square corner. A right angle measures exactly 90° and is typically indicated by a box drawn in the interior of the angle. An acute angle has an interior that is narrower than a right angle. The measure of an acute angle is any value less than 90° and greater than 0°. For example, 89.9°, 47°, 12°, and 1°. An obtuse angle has an interior that is wider than a right angle. The measure of an obtuse angle is any value greater than 90° but less than 180°. For example, 90.1°, 110°, 150°, and 179.9°.

- Acute angles: Less than 90°
- Obtuse angles: Greater than 90°
- Right angles: 90°
- Straight angles: 180°

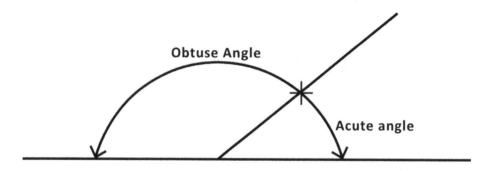

Adjacent angles are two side-by-side angles formed from the same ray that have the same endpoint.

Polygons

A *polygon* is a closed figure (meaning it divides the plane into an inside and an outside) consisting of a collection of line segments between points. These points are called the *vertices* of the polygon. These line segments must not overlap one another. Note that the number of sides is equal to the number of angles, or vertices of the polygon. The angles between line segments meeting one another in the polygon are called *interior angles*.

A *regular polygon* is a polygon whose edges are all the same length and whose interior angles are all of equal measure.

Polygons can be classified by the number of sides (also equal to the number of angles) they have. The following are the names of polygons with a given number of sides or angles:

# of sides	3	4	5	6	7	8	9	10
Name of polygon	Triangle	Quadrilateral	Pentagon	Hexagon	Septagon (or heptagon)	Octagon	Nonagon	Decagon

A *triangle* is a polygon with three sides. A *quadrilateral* is a polygon with four sides.

Quadrilaterals can be further classified according to their sides and angles. A quadrilateral with exactly one pair of parallel sides is called a trapezoid. A quadrilateral that shows both pairs of opposite sides parallel is a parallelogram. Parallelograms include rhombuses, rectangles, and squares. A rhombus has four equal sides. A rectangle has four equal angles (90° each). A square has four 90° angles and four equal sides. Therefore, a square is both a rhombus and a rectangle.

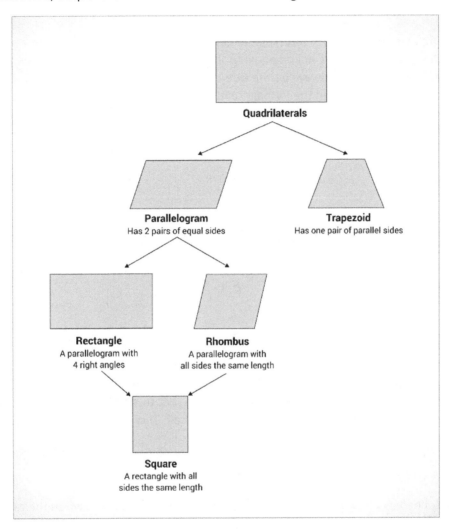

Triangles

A *right triangle* is a triangle that has one 90° angle.

The sum of the interior angles of any triangle must add up to 180°.

An *isosceles triangle* is a triangle in which two of the sides are the same length. In this case, it will always have two congruent interior angles. If a triangle has two congruent interior angles, it will always be isosceles.

An *equilateral triangle* is a triangle whose sides are all the same length and whose angles are all equivalent to one another, equal to 60°. Equilateral triangles are examples of regular polygons. Note that equilateral triangles are also isosceles. A triangle with no equal sides or angles is a scalene triangle.

Angles that add up to 90 degrees are *complementary*. Within a right triangle, two complementary angles exist because the third angle is always 90 degrees.

Pythagorean Theorem

The Pythagorean theorem is an important result in geometry. It states that for right triangles, the sum of the squares of the two shorter sides will be equal to the square of the longest side (also called the *hypotenuse*). The longest side will always be the side opposite to the 90° angle. If this side is called c, and the other two sides are a and b, then the Pythagorean theorem states that $c^2 = a^2 + b^2$. Since lengths are always positive, this also can be written as $c = \sqrt{a^2 + b^2}$.

A diagram to show the parts of a triangle using the Pythagorean theorem is below.

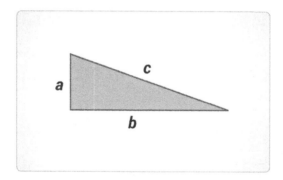

As an example of the theorem, suppose that Shirley has a rectangular field that is 5 feet wide and 12 feet long, and she wants to split it in half using a fence that goes from one corner to the opposite corner. How long will this fence need to be? To figure this out, note that this makes the field into two right triangles, whose hypotenuse will be the fence dividing it in half. Therefore, the fence length will be given by $\sqrt{5^2 + 12^2} = \sqrt{169} = 13$ feet long.

Circles

A *circle* can be defined as the set of all points that are the same distance (known as the radius, **r**) from a single point (known as the center of the circle). The center has coordinates (h, k), and any point on the circle can be labelled with coordinates (x, y).

The *circumference* of a circle is the distance traveled by following the edge of the circle for one complete revolution, and the length of the circumference is given by $2\pi r$, where r is the radius of the circle. The formula for circumference is $C = 2\pi r$.

Given two points on the circumference of a circle, the path along the circle between those points is called an *arc* of the circle. For example, the arc between *B* and *C* is denoted by a thinner line:

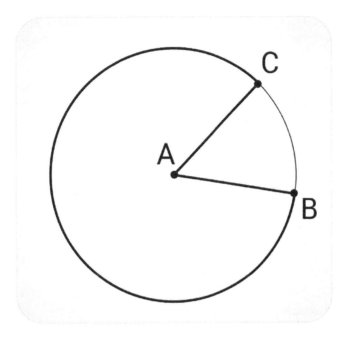

The length of the path along an arc is called the *arc length*. If the circle has radius *r*, then the arc length is given by multiplying the measure of the angle in radians by the radius of the circle.

The equation used to find the area of a circle is:

$$A = \pi r^2$$

For example, if a circle has a radius of 5 centimeters, the area is computed by substituting 5 for the radius: $(5)^2$. Using this reasoning, to find half of the area of a circle, the formula is:

$$A = .5\pi r^2$$

Similarly, to find the quarter of an area of a circle, the formula is:

$$A = .25\pi r^2$$

To find any fractional area of a circle, a student can use the formula $A = \frac{C}{360}\pi r^2$, where *C* is the number of degrees of the central angle of the sector. The area of a circle can also be found by using the arc length rather than the degree of the sector. This formula is $A = rs^2$, where *s* is the arc length and *r* is the radius of the circle.

Area and Perimeter

The perimeter of a polygon is the distance around the outside of the two-dimensional figure. Perimeter is a one-dimensional measurement and is therefore expressed in linear units such as centimeters (*cm*), feet (*ft*), and miles (*mi*). The perimeter (*P*) of a figure can be calculated by adding together each of the sides.

Properties of certain polygons allow that the perimeter may be obtained by using formulas. A rectangle consists of two sides called the length (l), which have equal measures, and two sides called the width (w), which have equal measures. Therefore, the perimeter (P) of a rectangle can be expressed as $P = l + l + w + w$. This can be simplified to produce the following formula to find the perimeter of a rectangle:

$$P = 2l + 2w \text{ or } P = 2(l + w)$$

A regular polygon is one in which all sides have equal length and all interior angles have equal measures, such as a square and an equilateral triangle. To find the perimeter of a regular polygon, the length of one side is multiplied by the number of sides. For example, to find the perimeter of an equilateral triangle with a side of length of 4 feet, 4 feet is multiplied by 3 (number of sides of a triangle). The perimeter of a regular octagon (8 sides) with a side of length of $\frac{1}{2}$cm is $\frac{1}{2}cm \times 8 = 4cm$.

The area of a polygon is the number of square units needed to cover the interior region of the figure. Area is a two-dimensional measurement. Therefore, area is expressed in square units, such as square centimeters (cm^2), square feet (ft^2), or square miles (mi^2). Regarding the area of a rectangle with sides of length x and y, the area is given by xy. For a triangle with a base of length b and a height of length h, the area is $\frac{1}{2}bh$. To find the area (A) of a parallelogram, the length of the base (b) is multiplied by the length of the height (h) → $A = b \times h$. Similar to triangles, the height of the parallelogram is measured from one base to the other at a 90° angle (or perpendicular).

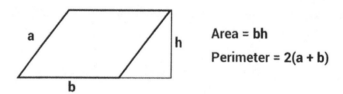

Area = bh

Perimeter = 2(a + b)

The area of a trapezoid can be calculated using the formula: $A = \frac{1}{2} \times h(b_1 + b_2)$, where h is the height and b_1 and b_2 are the parallel bases of the trapezoid.

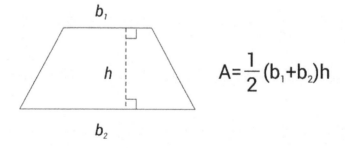

$$A = \frac{1}{2}(b_1 + b_2)h$$

The area of a regular polygon can be determined by using its perimeter and the length of the apothem. The apothem is a line from the center of the regular polygon to any of its sides at a right angle. (Note that the perimeter of a regular polygon can be determined given the length of only one side.) The

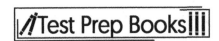

formula for the area (A) of a regular polygon is $A = \frac{1}{2} \times a \times P$, where a is the length of the apothem, and P is the perimeter of the figure.

Consider the following regular pentagon:

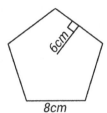

To find the area, the perimeter (P) is calculated first: $8cm \times 5 \rightarrow P = 40cm$. Then the perimeter and the apothem are used to find the area (A):

$$A = \frac{1}{2} \times a \times P$$

$$A = \frac{1}{2} \times (6cm) \times (40cm) \rightarrow A = 120cm^2$$

Note that the unit is:

$$cm^2 \rightarrow cm \times cm = cm^2$$

The area of irregular polygons is found by decomposing, or breaking apart, the figure into smaller shapes. When the area of the smaller shapes is determined, the area of the smaller shapes will produce the area of the original figure when added together. Consider the example below:

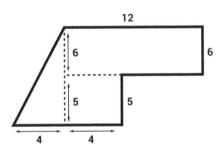

The irregular polygon is decomposed into two rectangles and a triangle. The area of the large rectangles ($A = l \times w \rightarrow A = 12 \times 6$) is 72 square units. The area of the small rectangle is 20 square units ($A = 4 \times 5$). The area of the triangle:

$$(A = \frac{1}{2} \times b \times h \rightarrow A = \frac{1}{2} \times 4 \times 11)$$

is 22 square units. The sum of the areas of these figures produces the total area of the original polygon:

$$A = 72 + 20 + 22 \rightarrow A = 114 \text{ square units}$$

Surface Area and Volume

Geometry in three dimensions is similar to geometry in two dimensions. The main new feature is that three points now define a unique *plane* that passes through each of them. Three dimensional objects can be made by putting together two dimensional figures in different surfaces. Below, some of the possible three dimensional figures will be provided, along with formulas for their volumes and surface areas.

Volume is the capacity of a three-dimensional shape. Volume is useful in determining the space within a certain three-dimensional object. Volume can be calculated for a cube, rectangular prism, cylinder, pyramid, cone, and sphere. By knowing specific dimensions of the objects, the volume of the object is computed with these figures. The units for the volumes of solids can include cubic centimeters, cubic meters, cubic inches, and cubic feet.

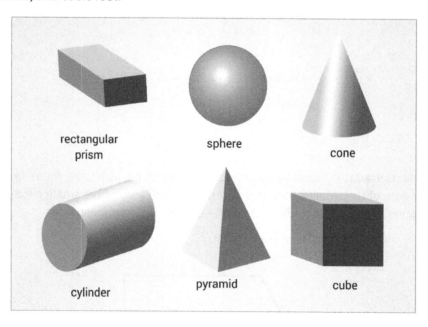

A rectangular prism is a box whose sides are all rectangles meeting at 90° angles. Such a box has three dimensions: length, width, and height. If the length is x, the width is y, and the height is z, then the volume is given by $V = xyz$.

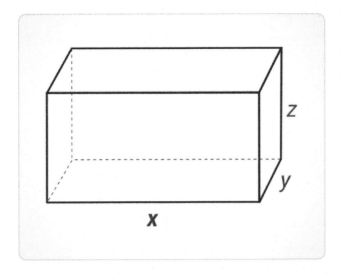

The surface area will be given by computing the surface area of each rectangle and adding them together. There are a total of six rectangles. Two of them have sides of length x and y, two have sides of length y and z, and two have sides of length x and z. Therefore, the total surface area will be given by:

$$SA = 2xy + 2yz + 2xz$$

A *cube* is a special type of rectangular solid in which its length, width, and height are the same. If this length is s, then the formula for the volume of a cube is:

$$V = s \times s \times s$$

The surface area of a cube is $SA = 6s^2$.

A *rectangular pyramid* is a figure with a rectangular base and four triangular sides that meet at a single vertex. If the rectangle has sides of length x and y, then the volume will be given by $V = \frac{1}{3}xyh$.

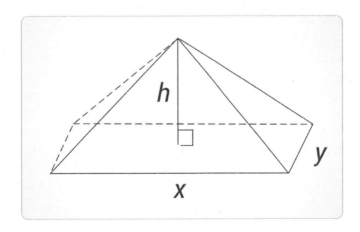

To find the surface area, the dimensions of each triangle need to be known. However, these dimensions can differ depending on the problem in question. Therefore, there is no general formula for calculating total surface area.

A *sphere* is a set of points all of which are equidistant from some central point. It is like a circle, but in three dimensions. The volume of a sphere of radius r is given by $V = \frac{4}{3}\pi r^3$. The surface area is given by $A = 4\pi r^2$.

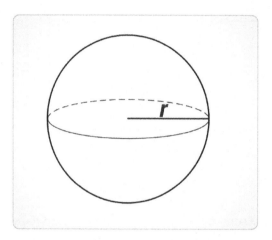

Discovering a cylinder's volume requires the measurement of the cylinder's base, length of the radius, and height. The height of the cylinder can be represented with variable h, and the radius can be represented with variable r.

The formula to find the volume of a cylinder is $\pi r^2 h$. Notice that πr^2 is the formula for the area of a circle. This is because the base of the cylinder is a circle. To calculate the volume of a cylinder, the slices of circles needed to build the entire height of the cylinder are added together. For example, if the radius is 5 feet and the height of the cylinder is 10 feet, the cylinder's volume is calculated by using the following equation: $\pi 5^2 \times 10$. Substituting 3.14 for π, the volume is 785.4 ft³.

Measurement

Converting Within and Between Standard and Metric Systems

American Measuring System
The measuring system used today in the United States developed from the British units of measurement during colonial times. The most typically used units in this customary system are those used to measure weight, liquid volume, and length, whose common units are found below. In the customary system, the basic unit for measuring weight is the ounce (oz); there are 16 ounces (oz) in 1 pound (lb) and 2000 pounds in 1 ton. The basic unit for measuring liquid volume is the ounce (oz); 1 ounce is equal to 2 tablespoons (tbsp) or 6 teaspoons (tsp), and there are 8 ounces in 1 cup, 2 cups in 1 pint (pt), 2 pints in 1 quart (qt), and 4 quarts in 1 gallon (gal). For measurements of length, the inch (in) is the base unit; 12 inches make up 1 foot (ft), 3 feet make up 1 yard (yd), and 5280 feet make up 1 mile (mi). However, as

there are only a set number of units in the customary system, with extremely large or extremely small amounts of material, the numbers can become awkward and difficult to compare.

Common Customary Measurements		
Length	**Weight**	**Capacity**
1 foot = 12 inches	1 pound = 16 ounces	1 cup = 8 fluid ounces
1 yard = 3 feet	1 ton = 2,000 pounds	1 pint = 2 cups
1 yard = 36 inches		1 quart = 2 pints
1 mile = 1,760 yards		1 quart = 4 cups
1 mile = 5,280 feet		1 gallon = 4 quarts
		1 gallon = 16 cups

Metric System

Aside from the United States, most countries in the world have adopted the metric system embodied in the International System of Units (SI). The three main SI base units used in the metric system are the meter (m), the kilogram (kg), and the liter (L); meters measure length, kilograms measure mass, and liters measure volume.

These three units can use different prefixes, which indicate larger or smaller versions of the unit by powers of ten. This can be thought of as making a new unit which is sized by multiplying the original unit in size by a factor.

These prefixes and associated factors are:

Metric Prefixes			
Prefix	**Symbol**	**Multiplier**	**Exponential**
kilo	k	1,000	10^3
hecto	h	100	10^2
deca	da	10	10^1
no prefix		1	10^0
deci	d	0.1	10^{-1}
centi	c	0.01	10^{-2}
milli	m	0.001	10^{-3}

The correct prefix is then attached to the base. Some examples:

1 milliliter equals .001 liters.

1 kilogram equals 1,000 grams.

Choosing the Appropriate Measuring Unit

Some units of measure are represented as square or cubic units depending on the solution. For example, perimeter is measured in units, area is measured in square units, and volume is measured in cubic units.

Also be sure to use the most appropriate unit for the thing being measured. A building's height might be measured in feet or meters while the length of a nail might be measured in inches or centimeters. Additionally, for SI units, the prefix should be chosen to provide the most succinct available value. For example, the mass of a bag of fruit would likely be measured in kilograms rather than grams or milligrams, and the length of a bacteria cell would likely be measured in micrometers rather than centimeters or kilometers.

Conversion

Converting measurements in different units between the two systems can be difficult because they follow different rules. The best method is to look up an English to Metric system conversion factor and then use a series of equivalent fractions to set up an equation to convert the units of one of the measurements into those of the other. The table below lists some common conversion values that are useful for problems involving measurements with units in both systems:

English System	Metric System
1 inch	2.54 cm
1 foot	0.3048 m
1 yard	0.914 m
1 mile	1.609 km
1 ounce	28.35 g
1 pound	0.454 kg
1 fluid ounce	29.574 mL
1 quart	0.946 L
1 gallon	3.785 L

Consider the example where a scientist wants to convert 6.8 inches to centimeters. The table above is used to find that there are 2.54 centimeters in every inch, so the following equation should be set up and solved:

$$\frac{6.8\ in}{1} \times \frac{2.54\ cm}{1\ in} = 17.272\ cm$$

Notice how the inches in the numerator of the initial figure and the denominator of the conversion factor cancel out. (This equation could have been written simply as $6.8\ in \times 2.54\ cm = 17.272\ cm$, but it was shown in detail to illustrate the steps). The goal in any conversion equation is to set up the fractions so that the units you are trying to convert from cancel out and the units you desire remain.

For a more complicated example, consider converting 2.15 kilograms into ounces. The first step is to convert kilograms into grams and then grams into ounces. Note that the measurement you begin with does not have to be put in a fraction.

So in this case, 2.15 kg is by itself although it's technically the numerator of a fraction:

$$2.15\ kg \times \frac{1000g}{kg} = 2150\ g$$

Then, use the conversion factor from the table to convert grams to ounces:

$$2150g \times \frac{1\ oz}{28.35g} = 75.8\ oz$$

Estimation

Rounding

Rounding numbers changes the given number to a simpler and less accurate number than the exact given number. Rounding allows for easier calculations which estimate the results of using the exact given number. The accuracy of the estimate and ease of use depends on the place value to which the number is rounded. Rounding numbers consists of:

- Determining what place value the number is being rounded to
- Examining the digit to the right of the desired place value to decide whether to round up or keep the digit
- Replacing all digits to the right of the desired place value with zeros

To round 746,311 to the nearest ten thousands, the digit in the ten thousands place should be located first. In this case, this digit is 4 (746,311). Then, the digit to its right is examined. If this digit is 5 or greater, the number will be rounded up by increasing the digit in the desired place by one. If the digit to the right of the place value being rounded is 4 or less, the number will be kept the same. For the given example, the digit being examined is a 6, which means that the number will be rounded up by increasing the digit to the left by one. Therefore, the digit 4 is changed to a 5. Finally, to write the rounded number, any digits to the left of the place value being rounded remain the same and any to its right are replaced with zeros. For the given example, rounding 746,311 to the nearest ten thousand will produce 750,000. To round 746,311 to the nearest hundred, the digit to the right of the three in the hundreds place is examined to determine whether to round up or keep the same number. In this case, that digit is a one, so the number will be kept the same and any digits to its right will be replaced with zeros. The resulting rounded number is 746,300.

Rounding place values to the right of the decimal follows the same procedure, but digits being replaced by zeros can simply be dropped. To round 3.752891 to the nearest thousandth, the desired place value is located (3.752891) and the digit to the right is examined. In this case, the digit 8 indicates that the number will be rounded up, and the 2 in the thousandths place will increase to a 3. Rounding up and replacing the digits to the right of the thousandths place produces 3.753000 which is equivalent to 3.753. Therefore, the zeros are not necessary and the rounded number should be written as 3.753.

When rounding up, if the digit to be increased is a 9, the digit to its left is increased by 1 and the digit in the desired place value is changed to a zero. For example, the number 1,598 rounded to the nearest ten is 1,600. Another example shows the number 43.72961 rounded to the nearest thousandth is 43.730 or 43.73.

Applying Estimation Strategies and Rounding Rules to Real-World Problems

Estimation

Estimation is finding a value that is close to a solution but is not the exact answer. For example, if there are values in the thousands to be multiplied, then each value can be estimated to the nearest thousand and the calculation performed. This value provides an approximate solution that can be determined very quickly.

Rounding Numbers

It's often convenient to round a number, which means to give an approximate figure to make it easier to compare amounts or perform mental math. Round up when the digit is 5 or more. The digit used to determine the rounding, and all subsequent digits, become 0, and the selected place value is increased by 1. Here are some examples:

75 rounded to the nearest ten is 80

380 rounded to the nearest hundred is 400

22.697 rounded to the nearest hundredth is 22.70

Round down when rounding on any digit that is below 5. The rounded digit, and all subsequent digits, becomes 0, and the preceding digit goes down by 1. Here are some examples:

92 rounded to the nearest ten is 90

839 rounded to the nearest hundred is 800

22.643 rounded to the nearest hundredth is 22.64

Determining the Reasonableness of Results

When solving math word problems, the solution obtained should make sense within the given scenario. The step of checking the solution will reduce the possibility of a calculation error or a solution that may be *mathematically* correct but not applicable in the real world. Consider the following scenarios:

A problem states that Lisa got 24 out of 32 questions correct on a test and asks to find the percentage of correct answers. To solve the problem, a student divided 32 by 24 to get 1.33, and then multiplied by 100 to get 133 percent. By examining the solution within the context of the problem, the student should recognize that getting all 32 questions correct will produce a perfect score of 100 percent. Therefore, a score of 133 percent with 8 incorrect answers does not make sense and the calculations should be checked.

A problem states that the maximum weight on a bridge cannot exceed 22,000 pounds. The problem asks to find the maximum number of cars that can be on the bridge at one time if each car weighs 4,000 pounds. To solve this problem, a student divided 22,000 by 4,000 to get an answer of 5.5. By examining the solution within the context of the problem, the student should recognize that although the calculations are mathematically correct, the solution does not make sense. Half of a car on a bridge is not possible, so the student should determine that a maximum of 5 cars can be on the bridge at the same time.

Mental Math Estimation

Once a result is determined to be logical within the context of a given problem, the result should be evaluated by its nearness to the expected answer. This is performed by approximating given values to perform mental math. Numbers should be rounded to the nearest value possible to check the initial results.

Consider the following example: A problem states that a customer is buying a new sound system for their home. The customer purchases a stereo for $435, 2 speakers for $67 each, and the necessary cables for $12. The customer chooses an option that allows him to spread the costs over equal payments for 4 months. How much will the monthly payments be?

After making calculations for the problem, a student determines that the monthly payment will be $145.25. To check the accuracy of the results, the student rounds each cost to the nearest ten ($440 + 70 + 70 + 10$) and determines that the total is approximately $590. Dividing by 4 months gives an approximate monthly payment of $147.50. Therefore, the student can conclude that the solution of $145.25 is very close to what should be expected.

When rounding, the place-value that is used in rounding can make a difference. Suppose the student had rounded to the nearest hundred for the estimation. The result ($400 + 100 + 100 + 0 = 600; 600 \div 4 = 150$) will show that the answer is reasonable, but not as close to the actual value as rounding to the nearest ten.

Precision and Accuracy

Precision and accuracy are used to describe groups of measurements. *Precision* describes a group of measures that are very close together, regardless of whether the measures are close to the true value. *Accuracy* describes how close the measures are to the true value.

Since accuracy refers to the closeness of a value to the true measurement, the level of accuracy depends on the object measured and the instrument used to measure it. This will vary depending on the situation. If measuring the mass of a set of dictionaries, kilograms may be used as the units. In this case, it is not vitally important to have a high level of accuracy. If the measurement is a few grams away from the true value, the discrepancy might not make a big difference in the problem.

In a different situation, the level of accuracy may be more significant. Pharmacists need to be sure they are very accurate in their measurements of medicines that they give to patients. In this case, the level of accuracy is vitally important and not something to be estimated. In the dictionary situation, the measurements were given as whole numbers in kilograms. In the pharmacist's situation, the measurements for medicine must be taken to the milligram and sometimes further, depending on the type of medicine.

When considering the accuracy of measurements, the error in each measurement can be shown as absolute and relative. *Absolute error* tells the actual difference between the measured value and the true value. The *relative error* tells how large the error is in relation to the true value. There may be two problems where the absolute error of the measurements is 10 grams. For one problem, this may mean the relative error is very small because the measured value is 14,990 grams, and the true value is 15,000 grams. Ten grams in relation to the true value of 15,000 is small: 0.06%. For the other problem, the measured value is 290 grams, and the true value is 300 grams. In this case, the 10-gram absolute error means a high relative error because the true value is smaller. The relative error is 10/300 = 0.03, or 3%.

135

Practice Questions

Concepts, Data Interpretation, and Problem Solving

1. The graph of which function has an x-intercept of -2?
   a. $y = 2x - 3$
   b. $y = 4x + 2$
   c. $y = x^2 + 5x + 6$
   d. $y = -\frac{1}{2} \times 2^x$

2. The table below displays the number of three-year-olds at Kids First Daycare who are potty-trained and those who still wear diapers.

	Potty-trained	Wear diapers	
Boys	26	22	48
Girls	34	18	52
	60	40	

What is the probability that a three-year-old girl chosen at random from the school is potty-trained?
   a. 52 percent
   b. 34 percent
   c. 65 percent
   d. 57 percent

3. Express the solution to the following problem in decimal form:
$$\frac{3}{5} \times \frac{7}{10} \div \frac{1}{2}$$

   a. 0.042
   b. 84%
   c. 0.84
   d. 0.42

4. Karen gets paid a weekly salary and a commission for every sale that she makes. The table below shows the number of sales and her pay for different weeks.

Sales	2	7	4	8
Pay	$380	$580	$460	$620

Which of the following equations represents Karen's weekly pay?
   a. $y = 90x + 200$
   b. $y = 90x - 200$
   c. $y = 40x + 300$
   d. $y = 40x - 300$

5. What are the y-intercept(s) for $y = x^2 + 3x - 4$?

    a. $y = 1$

    b. $y = -4$

    c. $y = 3$

    d. $y = 4$

Estimation

Directions: Estimate the answer in your head. No writing is permitted. An exact answer is not expected.

1. If a car can travel 300 miles in 4 hours, about how far can it go in an hour and a half?

    a. 100 miles

    b. 110 miles

    c. 125 miles

    d. 150 miles

2. A traveler takes an hour to drive to a museum, spends 3 hours and 30 minutes there, and takes half an hour to drive home. Around what percentage of his or her time was spent driving?

    a. 15%

    b. 30%

    c. 40%

    d. 60%

3. What is the approximate integer equivalent of $\frac{660}{100}$?

    a. 67

    b. 66

    c. 7

    d. 6

4. Johnny is a competitive runner. He knows that a 5k run is 3.1 miles. About how many kilometers does he need to run if he wants to run roughly 15 miles?

    a. 9 km

    b. 18 km

    c. 21 km

    d. 25 km

5. Karina wants to adopt a puppy from the shelter. The adoption fee is $45. It costs $25 to register the dog, and the initial vet appointment will be $250. Lastly, the initial food, toys, and bedding will cost approximately $90. She earns $11 per hour babysitting. About how many hours does she need to work to cover the upfront costs to adopt her dog?

    a. 20 hours

    b. 30 hours

    c. 40 hours

    d. 50 hours

Answer Explanations

Concepts, Data Interpretation, and Problem Solving

1. C: An *x*-intercept is the point where the graph crosses the *x*-axis. At this point, the value of *y* is 0. To determine if an equation has an *x*-intercept of −2, substitute −2 for *x*, and calculate the value of *y*. If the value of −2 for *x* corresponds with a *y*-value of 0, then the equation has an *x*-intercept of −2. The only answer choice that produces this result is:

$$\text{Choice } C \rightarrow 0 = (-2)2 + 5(-2) + 6$$

2. C: The conditional frequency of a girl being potty-trained is calculated by dividing the number of potty-trained girls by the total number of girls: $34 \div 52 = 0.65$. To determine the conditional probability, multiply the conditional frequency by 100: $0.65 \times 100 = 65\%$.

3. C: The first step in solving this problem is expressing the result in fraction form. Separate this problem first by solving the division operation of the last two fractions. When dividing one fraction by another, invert or flip the second fraction and then multiply the numerator and denominator.

$$\frac{7}{10} \times \frac{2}{1} = \frac{14}{10}$$

Next, multiply the first fraction with this value:

$$\frac{3}{5} \times \frac{14}{10} = \frac{42}{50}$$

Decimals are expressions of 1 or 100%, so multiply both the numerator and denominator by 2 to get the fraction as an expression of 100.

$$\frac{42}{50} \times \frac{2}{2} = \frac{84}{100}$$

In decimal form, this would be expressed as 0.84.

4. C: $y = 40x + 300$

In this scenario, the variables are the number of sales and Karen's weekly pay. The weekly pay depends on the number of sales. Therefore, weekly pay is the dependent variable (*y*), and the number of sales is the independent variable (*x*). Each pair of values from the table can be written as an ordered pair (*x, y*): (2, 380), (7, 580), (4, 460), (8, 620).

The ordered pairs can be substituted into the equations to see which creates true statements (both sides equal) for each pair. Even if one ordered pair produces equal values for a given equation, the other three ordered pairs must be checked. The only equation which is true for all four ordered pairs is

$$y = 40x + 300$$

$$380 = 40(2) + 300 \rightarrow 380 = 380$$

$$580 = 40(7) + 300 \rightarrow 580 = 580$$

$$460 = 40(4) + 300 \rightarrow 460 = 460$$

$$620 = 40(8) + 300 \rightarrow 620 = 620$$

5. B: The y-intercept of an equation is found where the x-value is zero. Plugging zero into the equation for x, the first two terms cancel out, leaving -4.

Estimation

1. B: 300 miles in 4 hours is 300/4 = 75 miles per hour. In 1.5 hours, the car will go 1.5 $\times$ 75 miles, or, which can be estimated to be 75 + 35 or 110 miles.

2. B: The total trip time is 1 + 3.5 + 0.5 = 5 hours. The total time driving is 1 + 0.5 = 1.5 hours. So, the fraction of time spent driving is 1.5/5 or 3/10. To get the percentage, convert this to a fraction out of 100. The numerator and denominator are multiplied by 10, with a result of 30/100. The percentage is the numerator in a fraction out of 100, so 30%.

3. C: Dividing by 100 means shifting the decimal point of the numerator to the left by 2. The result is 6.6 and rounds to 7.

4. D: Because there are about 5km in 3 miles, if he runs 15 miles, which is 3 miles x 5, he runs 5 km x 5, or 25 km.

5. C: First, we need to approximate the initial costs. 50 + 25 + 250 + 100 = $425. If she makes 11/hr, we can use 10/hr for our estimation. 425/10 = about 40 hours. Because we rounded down on her wage and rounded up on the expenses, 40 hours is a good estimation.

Ability

The Abilities section of the TACHS assesses the test taker's reasoning abilities. Unlike questions in the other sections of the exam, which mostly cover content students will have covered in their formal schooling, the questions on the Abilities section require knowledge and skill development from out-of-school life experiences as well. For example, many questions use spatial and pattern reasoning presented in a novel way that may be unfamiliar to test takers. This is intentional because it helps the test evaluators gauge the test taker's ability to reason and solve problems that can be logically worked through even though the test taker may not have been directly taught the material.

Test takers are allotted thirty-two minutes to complete the questions in the Abilities section. These questions are broken down into two parts. The first part, Similarities and Changes, consists of 40 questions that must be completed in 25 minutes. In this section, test takers will encounter questions designed to assess their ability to understand and analyze sequences and patterns. The second section, Abstract Reasoning, contains only 10 questions, which must be completed in seven minutes. These questions require test takers to make predictions and inferences about patterns they are presented.

There are a variety of different patterns and exercises that students might encounter on the Abilities section, which all require test takers to use their creativity, logic, and spatial reasoning skills. The following three activities are examples of exercises found in the Abilities section of the TACHS:

1. Figure Matrices: Matrices are presented that consist of different figures that are somehow related and adhere to specific rules. The figures found in every row for example, adhere to one rule. Figures found in each column adhere to another rule. Test takers must examine the figures in the matrix and determine these implicit rules and the ways in which the figures and the rows and columns are related. Then, the test taker must select the figure from the available choices that correctly fills the missing space, given the relations and rules he or she inferred. For example, consider a pinwheel shape with six different petals. In each successive row, an additional petal might be added to the vertical stem going in the clockwise direction starting at the 12:00 position. In the columns direction, moving counterclockwise around the pinwheel, the outlined petals might be filled in black one at a time per successive column. If the missing figure is the one that needs to occupy the fourth row and second column, it would have four petals starting at the 12:00 position and moving towards 4:00. The last two (and 3:00 and 4:00) would be filled in black.

2. Paper Folding: A piece of paper is folded in some configuration (in half lengthwise, in half diagonally, in quarters, etc.) and then punched with a hole punch. Test takers must imagine where the holes would end up when the paper would be fully unfolded. They must select the answer choice that correctly depicts the location and number of expected holes on the unfolded paper. For example, if the prompt shows a paper folded diagonally with two holes created along the crease—one on either end—the unfolded paper would have four holes located near two opposite corners straddling the former diagonal crease. Choices that have fewer or more than four holes would be incorrect, as would any answer choices that positioned the four holes in an alternative configuration on the unfolded paper (near the direct center of the paper, for instance).

3. Figure Classification: Each question contains three figures that are somehow related to one another. Test takers must determine this relationship and then select the answer choice that

maintains this relationship and fits with the initial three images. For example, three triangles shaded in with horizontal gray stripes may be depicted in the question prompt. The sizes of the three triangles may all be different (for example, a very small triangle, a large one, and a medium-sized triangle), but their orientation, type (for example, isosceles), and appearance (in terms of the shading pattern) will all be identical. The correct answer choice will maintain these similarities, so it will also be an isosceles triangle that is shaded with horizontal gray stripes. The size will be the only factor that may vary, meaning it might be larger, smaller, or somewhere in the mid-range of the other three triangles.

Practice Questions

Figure Matrices

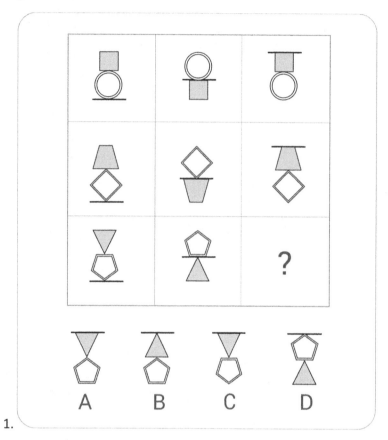

1.

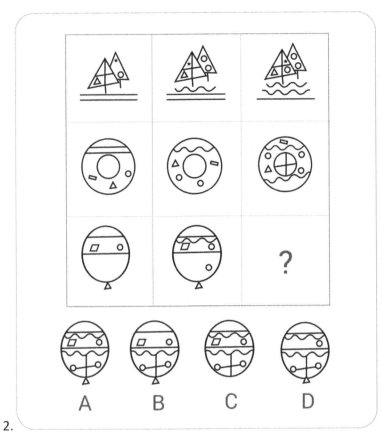

2.

A   B   C   D

3.

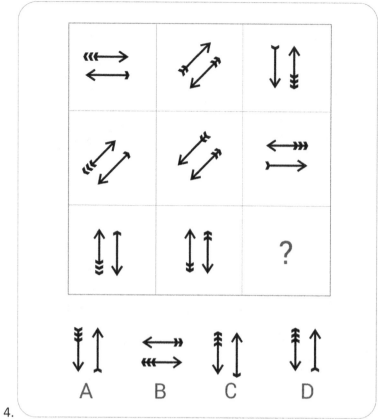

4.

5.

Paper Folding

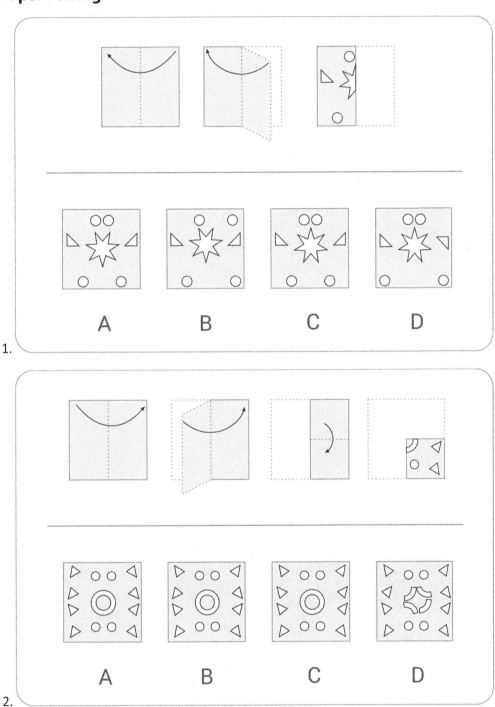

1.

2.

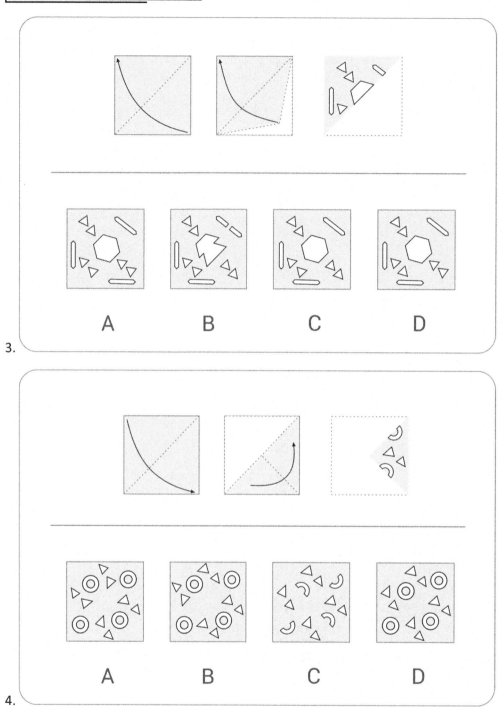

3.

4.

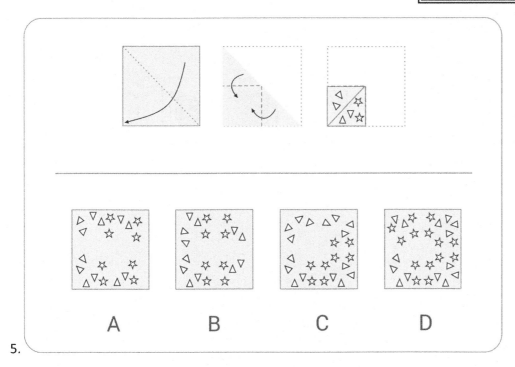

5.

Figure Classification

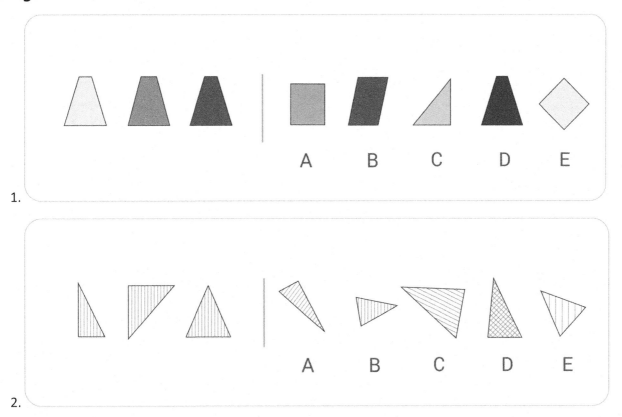

1.

2.

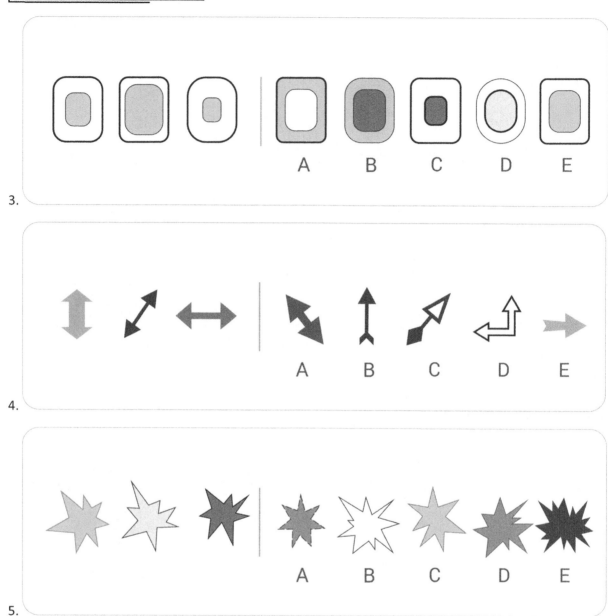

3.

4.

5.

Answer Explanations

Figure Matrices

1. C

2. A

3. B

4. D

5. C

Paper Folding

1. C

2. B

3. D

4. A

5. C

Figure Classification

1. D

2. B

3. E

4. A

5. C

Practice Questions

Reading

Passage #1

Questions 1-6 are based upon the following passage:

This excerpt is an adaptation of Jonathan Swift's *Gulliver's Travels into Several Remote Nations of the World.*

My gentleness and good behaviour had gained so far on the emperor and his court, and indeed upon the army and people in general, that I began to conceive hopes of getting my liberty in a short time. I took all possible methods to cultivate this favourable disposition. The natives came, by degrees, to be less apprehensive of any danger from me. I would sometimes lie down, and let five or six of them dance on my hand; and at last the boys and girls would venture to come and play at hide-and-seek in my hair. I had now made a good progress in understanding and speaking the language. The emperor had a mind one day to entertain me with several of the country shows, wherein they exceed all nations I have known, both for dexterity and magnificence. I was diverted with none so much as that of the rope-dancers, performed upon a slender white thread, extended about two feet, and twelve inches from the ground. Upon which I shall desire liberty, with the reader's patience, to enlarge a little.

This diversion is only practised by those persons who are candidates for great employments, and high favour at court. They are trained in this art from their youth, and are not always of noble birth, or liberal education. When a great office is vacant, either by death or disgrace (which often happens,) five or six of those candidates petition the emperor to entertain his majesty and the court with a dance on the rope; and whoever jumps the highest, without falling, succeeds in the office. Very often the chief ministers themselves are commanded to show their skill, and to convince the emperor that they have not lost their faculty. Flimnap, the treasurer, is allowed to cut a caper on the straight rope, at least an inch higher than any other lord in the whole empire. I have seen him do the summerset several times together, upon a trencher fixed on a rope which is no thicker than a common packthread in England. My friend Reldresal, principal secretary for private affairs, is, in my opinion, if I am not partial, the second after the treasurer; the rest of the great officers are much upon a par.

1. Which of the following statements best summarize the central purpose of this text?
    a. Gulliver details his fondness for the archaic yet interesting practices of his captors.
    b. Gulliver conjectures about the intentions of the aristocratic sector of society.
    c. Gulliver becomes acquainted with the people and practices of his new surroundings.
    d. Gulliver's differences cause him to become penitent around new acquaintances.

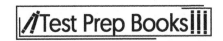

2. What is the word *principal* referring to in the following text?

> My friend Reldresal, principal secretary for private affairs, is, in my opinion, if I am not partial, the second after the treasurer; the rest of the great officers are much upon a par.

a. Primary or chief
b. An acolyte
c. An individual who provides nurturing
d. One in a subordinate position

3. What can the reader infer from this passage?

> I would sometimes lie down, and let five or six of them dance on my hand; and at last the boys and girls would venture to come and play at hide-and-seek in my hair.

a. The children tortured Gulliver.
b. Gulliver traveled because he wanted to meet new people.
c. Gulliver is considerably larger than the children who are playing around him.
d. Gulliver has a genuine love and enthusiasm for people of all sizes.

4. What is the significance of the word *mind* in the following passage?

> The emperor had a mind one day to entertain me with several of the country shows, wherein they exceed all nations I have known, both for dexterity and magnificence.

a. The ability to think
b. A collective vote
c. A definitive decision
d. A mythological question

5. Which of the following assertions does not support the fact that games are a commonplace event in this culture?

a. My gentlest and good behavior . . . short time.
b. They are trained in this art from their youth . . . liberal education.
c. Very often the chief ministers themselves are commanded to show their skill . . . not lost their faculty.
d. Flimnap, the treasurer, is allowed to cut a caper on the straight rope . . . higher than any other lord in the whole empire.

6. How do the roles of Flimnap and Reldresal serve as evidence of the community's emphasis in regards to the correlation between physical strength and leadership abilities?

a. Only children used Gulliver's hands as a playground.
b. The two men who exhibited superior abilities held prominent positions in the community.
c. Only common townspeople, not leaders, walk the straight rope.
d. No one could jump higher than Gulliver.

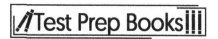

Passage #2

Questions 7-12 are based upon the following passage:

This excerpt is adaptation of Robert Louis Stevenson's *The Strange Case of Dr. Jekyll and Mr. Hyde.*

"Did you ever come across a protégé of his—one Hyde?" He asked.

"Hyde?" repeated Lanyon. "No. Never heard of him. Since my time."

That was the amount of information that the lawyer carried back with him to the great, dark bed on which he tossed to and fro until the small hours of the morning began to grow large. It was a night of little ease to his toiling mind, toiling in mere darkness and besieged by questions.

Six o'clock struck on the bells of the church that was so conveniently near to Mr. Utterson's dwelling, and still he was digging at the problem. Hitherto it had touched him on the intellectual side alone; but; but now his imagination also was engaged, or rather enslaved; and as he lay and tossed in the gross darkness of the night in the curtained room, Mr. Enfield's tale went by before his mind in a scroll of lighted pictures. He would be aware of the great field of lamps in a nocturnal city; then of the figure of a man walking swiftly; then of a child running from the doctor's; and then these met, and that human Juggernaut trod the child down and passed on regardless of her screams. Or else he would see a room in a rich house, where his friend lay asleep, dreaming and smiling at his dreams; and then the door of that room would be opened, the curtains of the bed plucked apart, the sleeper recalled, and, lo! There would stand by his side a figure to whom power was given, and even at that dead hour he must rise and do its bidding. The figure in these two phrases haunted the lawyer all night; and if at anytime he dozed over, it was but to see it glide more stealthily through sleeping houses, or move the more swiftly, and still the more smoothly, even to dizziness, through wider labyrinths of lamplighted city, and at every street corner crush a child and leave her screaming. And still the figure had no face by which he might know it; even in his dreams it had no face, or one that baffled him and melted before his eyes; and thus there it was that there sprung up and grew apace in the lawyer's mind a singularly strong, almost an inordinate, curiosity to behold the features of the real Mr. Hyde. If he could but once set eyes on him, he thought the mystery would lighten and perhaps roll altogether away, as was the habit of mysterious things when well examined. He might see a reason for his friend's strange preference or bondage, and even for the startling clauses of the will. And at least it would be a face worth seeing: the face of a man who was without bowels of mercy: a face which had but to show itself to raise up, in the mind of the unimpressionable Enfield, a spirit of enduring hatred.

From that time forward, Mr. Utterson began to haunt the door in the by street of shops. In the morning before office hours, at noon when business was plenty of time scarce, at night under the face of the full city moon, by all lights and at all hours of solitude or concourse, the lawyer was to be found on his chosen post.

"If he be Mr. Hyde," he had thought, "I should be Mr. Seek."

7. What is the purpose of the use of repetition in the following passage?

> It was a night of little ease to his toiling mind, toiling in mere darkness and besieged by questions.

    a. It serves as a demonstration of the mental state of Mr. Lanyon.
    b. It is reminiscent of the church bells that are mentioned in the story.
    c. It mimics Mr. Utterson's ambivalence.
    d. It emphasizes Mr. Utterson's anguish in failing to identify Hyde's whereabouts.

8. What is the setting of the story in this passage?
    a. In the city
    b. On the countryside
    c. In a jail
    d. In a mental health facility

9. What can one infer about the meaning of the word "Juggernaut" from the author's use of it in the passage?
    a. It is an apparition that appears at daybreak.
    b. It scares children.
    c. It is associated with space travel.
    d. Mr. Utterson finds it soothing.

10. What is the definition of the word *haunt* in the following passage?

> From that time forward, Mr. Utterson began to haunt the door in the by street of shops. In the morning before office hours, at noon when business was plenty of time scarce, at night under the face of the full city moon, by all lights and at all hours of solitude or concourse, the lawyer was to be found on his chosen post.

    a. To levitate
    b. To constantly visit
    c. To terrorize
    d. To daunt

11. The phrase *labyrinths of lamplighted city* contains an example of what?
    a. Hyperbole
    b. Simile
    c. Juxtaposition
    d. Alliteration

12. What can one reasonably conclude from the final comment of this passage?

> "If he be Mr. Hyde," he had thought, "I should be Mr. Seek."

    a. The speaker is considering a name change.
    b. The speaker is experiencing an identity crisis.
    c. The speaker has mistakenly been looking for the wrong person.
    d. The speaker intends to continue to look for Hyde.

Passage #3

Questions 13-18 are based upon the following passage:

This excerpt is adaptation from "What to the Slave is the Fourth of July?" Rochester, New York July 5, 1852

Fellow citizens—Pardon me, and allow me to ask, why am I called upon to speak here today? What have I, or those I represent, to do with your national independence? Are the great principles of political freedom and of natural justice embodied in that Declaration of Independence, Independence extended to us? And am I therefore called upon to bring our humble offering to the national altar, and to confess the benefits, and express devout gratitude for the blessings, resulting from your independence to us?

Would to God, both for your sakes and ours, ours that an affirmative answer could be truthfully returned to these questions! Then would my task be light, and my burden easy and delightful. For who is there so cold that a nation's sympathy could not warm him? Who so obdurate and dead to the claims of gratitude that would not thankfully acknowledge such priceless benefits? Who so stolid and selfish, that would not give his voice to swell the hallelujahs of a nation's jubilee, when the chains of servitude had been torn from his limbs? I am not that man. In a case like that, the dumb may eloquently speak, and the lame man leap as an hart.

But, such is not the state of the case. I say it with a sad sense of the disparity between us. I am not included within the pale of this glorious anniversary. Oh pity! Your high independence only reveals the immeasurable distance between us. The blessings in which you this day rejoice, I do not enjoy in common. The rich inheritance of justice, liberty, prosperity, and independence, bequeathed by your fathers, is shared by *you*, not by *me*. This Fourth of July is *yours,* not *mine.* You may rejoice, *I* must mourn. To drag a man in fetters into the grand illuminated temple of liberty, and call upon him to join you in joyous anthems, were inhuman mockery and sacrilegious irony. Do you mean, citizens, to mock me, by asking me to speak today? If so there is a parallel to your conduct. And let me warn you that it is dangerous to copy the example of a nation whose crimes, towering up to heaven, were thrown down by the breath of the Almighty, burying that nation and irrecoverable ruin! I can today take up the plaintive lament of a peeled and woe-smitten people.

By the rivers of Babylon, there we sat down. Yea! We wept when we remembered Zion. We hanged our harps upon the willows in the midst thereof. For there, they that carried us away captive, required of us a song; and they who wasted us required of us mirth, saying, "Sing us one of the songs of Zion." How can we sing the Lord's song in a strange land? If I forget thee, O Jerusalem, let my right hand forget her cunning. If I do not remember thee, let my tongue cleave to the roof of my mouth.

13. What is the tone of the first paragraph of this passage?
    a. Exasperated
    b. Inclusive
    c. Contemplative
    d. Nonchalant

14. Which word CANNOT be used synonymously with the term *obdurate* as it is conveyed in the text below?

> Who so obdurate and dead to the claims of gratitude, that would not thankfully acknowledge such priceless benefits?

a. Steadfast
b. Stubborn
c. Contented
d. Unwavering

15. What is the central purpose of this text?
a. To demonstrate the author's extensive knowledge of the Bible
b. To address the feelings of exclusion expressed by African Americans after the establishment of the Fourth of July holiday
c. To convince wealthy landowners to adopt new holiday rituals
d. To explain why minorities often relished the notion of segregation in government institutions

16. Which statement serves as evidence of the question above?
a. By the rivers of Babylon . . . down.
b. Fellow citizens . . . today.
c. I can . . . woe-smitten people.
d. The rich inheritance of justice . . . *not by me.*

17. The statement below features an example of which of the following literary devices?

> Oh pity! Your high independence only reveals the immeasurable distance between us.

a. Assonance
b. Parallelism
c. Amplification
d. Hyperbole

18. The speaker's use of biblical references, such as "rivers of Babylon" and the "songs of Zion," helps the reader to do all of the following EXCEPT:
a. Identify with the speaker through the use of common text.
b. Convince the audience that injustices have been committed by referencing another group of people who have been previously affected by slavery.
c. Display the equivocation of the speaker and those that he represents.
d. Appeal to the listener's sense of humanity.

Passage #4

Questions 19-24 are based upon the following passage:

This excerpt is an adaptation from Abraham Lincoln's Address Delivered at the Dedication of the Cemetery at Gettysburg, November 19, 1863.

> Four score and seven years ago our fathers brought forth on this continent, a new nation, conceived in liberty, and dedicated to the proposition that all men are created equal.
>
> Now we are engaged in a great civil war, testing whether that nation, or any nation so conceived and so dedicated, can long endure. We are met on a great battlefield of that war. We have come to dedicate a portion of that field, as a final resting place for those who here gave their lives that this nation might live. It is altogether fitting and proper that we should do this.
>
> But, in a larger sense, we cannot dedicate—we cannot consecrate that we cannot hallow—this ground. The brave men, living and dead, who struggled here, have consecrated it, far above our poor power to add or detract. The world will little note, nor long remember what we say here, but it can never forget what they did here. It is for us the living, rather, to be dedicated here to the unfinished work which they who fought here have thus far so nobly advanced. It is rather for us to be here and dedicated to the great task remaining before us—that from these honored dead we take increased devotion to that cause for which they gave the last full measure of devotion—that we here highly resolve that these dead shall not have died in vain—that these this nation, under God, shall have a new birth of freedom—and that government of people, by the people, for the people, shall not perish from the earth.

19. The best description for the phrase *four score and seven years ago* is which of the following?
    a. A unit of measurement
    b. A period of time
    c. A literary movement
    d. A statement of political reform

20. What is the setting of this text?
    a. A battleship off of the coast of France
    b. A desert plain on the Sahara Desert
    c. A battlefield in North America
    d. The residence of Abraham Lincoln

21. Which war is Abraham Lincoln referring to in the following passage?
    Now we are engaged in a great civil war, testing whether that nation, or any nation so conceived and so dedicated, can long endure.

    a. World War I
    b. The War of the Spanish Succession
    c. World War II
    d. The American Civil War

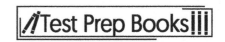

22. What message is the author trying to convey through this address?
    a. The audience should consider the death of the people that fought in the war as an example and perpetuate the ideals of freedom that the soldiers died fighting for.
    b. The audience should honor the dead by establishing an annual memorial service.
    c. The audience should form a militia that would overturn the current political structure.
    d. The audience should forget the lives that were lost and discredit the soldiers.

23. Which rhetorical device is being used in the following passage?
    . . . we here highly resolve that these dead shall not have died in vain—that these this nation, under God, shall have a new birth of freedom—and that government of people, by the people, for the people, shall not perish from the earth.

    a. Antimetabole
    b. Antiphrasis
    c. Anaphora
    d. Epiphora

24. What is the effect of Lincoln's statement in the following passage?
    But, in a larger sense, we cannot dedicate—we cannot consecrate that we cannot hallow—this ground. The brave men, living and dead, who struggled here, have consecrated it, far above our poor power to add or detract.

    a. His comparison emphasizes the great sacrifice of the soldiers who fought in the war.
    b. His comparison serves as a reminder of the inadequacies of his audience.
    c. His comparison serves as a catalyst for guilt and shame among audience members.
    d. His comparison attempts to illuminate the great differences between soldiers and civilians.

Passage #5

Questions 25-30 are based upon the following passage:

This excerpt is adaptation from Charles Dickens' speech in Birmingham in England on December 30, 1853 on behalf of the Birmingham and Midland Institute.

> My Good Friends,—When I first imparted to the committee of the projected Institute my particular wish that on one of the evenings of my readings here the main body of my audience should be composed of working men and their families, I was animated by two desires; first, by the wish to have the great pleasure of meeting you face to face at this Christmas time, and accompany you myself through one of my little Christmas books; and second, by the wish to have an opportunity of stating publicly in your presence, and in the presence of the committee, my earnest hope that the Institute will, from the beginning, recognise one great principle—strong in reason and justice—which I believe to be essential to the very life of such an Institution. It is, that the working man shall, from the first unto the last, have a share in the management of an Institution which is designed for his benefit, and which calls itself by his name.

> I have no fear here of being misunderstood—of being supposed to mean too much in this. If there ever was a time when any one class could of itself do much for its own good, and for the welfare of society—which I greatly doubt—that time is unquestionably past. It is in the fusion of different classes, without confusion; in the

157

bringing together of employers and employed; in the creating of a better common understanding among those whose interests are identical, who depend upon each other, who are vitally essential to each other, and who never can be in unnatural antagonism without deplorable results, that one of the chief principles of a Mechanics' Institution should consist. In this world a great deal of the bitterness among us arises from an imperfect understanding of one another. Erect in Birmingham a great Educational Institution, properly educational; educational of the feelings as well as of the reason; to which all orders of Birmingham men contribute; in which all orders of Birmingham men meet; wherein all orders of Birmingham men are faithfully represented—and you will erect a Temple of Concord here which will be a model edifice to the whole of England.

Contemplating as I do the existence of the Artisans' Committee, which not long ago considered the establishment of the Institute so sensibly, and supported it so heartily, I earnestly entreat the gentlemen—earnest I know in the good work, and who are now among us,—by all means to avoid the great shortcoming of similar institutions; and in asking the working man for his confidence, to set him the great example and give him theirs in return. You will judge for yourselves if I promise too much for the working man, when I say that he will stand by such an enterprise with the utmost of his patience, his perseverance, sense, and support; that I am sure he will need no charitable aid or condescending patronage; but will readily and cheerfully pay for the advantages which it confers; that he will prepare himself in individual cases where he feels that the adverse circumstances around him have rendered it necessary; in a word, that he will feel his responsibility like an honest man, and will most honestly and manfully discharge it. I now proceed to the pleasant task to which I assure you I have looked forward for a long time.

25. Which word is most closely synonymous with the word *patronage* as it appears in the following statement?
> . . . that I am sure he will need no charitable aid or condescending patronage

a. Auspices
b. Aberration
c . Acerbic
d. Adulation

26. Which term is most closely aligned with the definition of the term *working man* as it is defined in the following passage?
> You will judge for yourselves if I promise too much for the working man, when I say that he will stand by such an enterprise with the utmost of his patience, his perseverance, sense, and support . . .

a. Plebeian
b. Viscount
c. Entrepreneur
d. Bourgeois

27. Which of the following statements most closely correlates with the definition of the term *working man* as it is defined in Question 26?

    a. A working man is not someone who works for institutions or corporations, but someone who is well versed in the workings of the soul.

    b. A working man is someone who is probably not involved in social activities because the physical demand for work is too high.

    c. A working man is someone who works for wages among the middle class.

    d. The working man has historically taken to the field, to the factory, and now to the screen.

28. Based upon the contextual evidence provided in the passage above, what is the meaning of the term *enterprise* in the third paragraph?

    a. Company

    b. Courage

    c. Game

    d. Cause

29. The speaker addresses his audience as *My Good Friends*—what kind of credibility does this salutation give to the speaker?

    a. The speaker is an employer addressing his employees, so the salutation is a way for the boss to bridge the gap between himself and his employees.

    b. The speaker's salutation is one from an entertainer to his audience, and uses the friendly language to connect to his audience before a serious speech.

    c. The salutation gives the serious speech that follows a somber tone, as it is used ironically.

    d. The speech is one from a politician to the public, so the salutation is used to grab the audience's attention.

30. According to the aforementioned passage, what is the speaker's second desire for his time in front of the audience?

    a. To read a Christmas story

    b. For the working man to have a say in his institution which is designed for his benefit

    c. To have an opportunity to stand in their presence

    d. For the life of the institution to be essential to the audience as a whole

Written Expression

1. Which of the following sentences has an error in capitalization?

    a. The East Coast has experienced very unpredictable weather this year.

    b. My Uncle owns a home in Florida, where he lives in the winter.

    c. I am taking English Composition II on campus this fall.

    d. There are several nice beaches we can visit on our trip to the Jersey Shore this summer.

2. Which of the following sentences uses correct punctuation?

    a. Carole is not currently working; her focus is on her children at the moment.

    b. Carole is not currently working and her focus is on her children at the moment.

    c. Carole is not currently working, her focus is on her children at the moment.

    d. Carole is not currently working her focus is on her children at the moment.

3. All of Shannon's family and friends helped her to celebrate her 50th birthday at Café Sorrento. Which of the following is the complete subject of the preceding sentence?
   a. Family and friends
   b. All
   c. All of Shannon's family and friends
   d. Shannon's family and friends

4. Which of the following sentences correctly uses a hyphen?
   a. Last-year, many of the players felt unsure of the coach's methods.
   b. Some of the furniture she selected seemed a bit over - the - top for the space.
   c. Henry is a beagle-mix and is ready for adoption this weekend.
   d. Geena works to maintain a good relationship with her ex-husband to the benefit of their children.

5. Which of the following examples correctly uses quotation marks?
   a. "Where the Red Fern Grows" was one of my favorite novels as a child.
   b. Though he is famous for his roles in films like "The Great Gatsby" and "Titanic," Leonardo DiCaprio has never won an Oscar.
   c. Sylvia Plath's poem, "Daddy" will be the subject of this week's group discussion.
   d. "The New York Times" reported that many fans are disappointed in some of the trades made by the Yankees this off-season.

6. Which of the following sentences shows correct word usage?
   a. It's often been said that work is better then rest.
   b. Its often been said that work is better then rest.
   c. It's often been said that work is better than rest.
   d. Its often been said that work is better than rest.

7. Which of the following uses correct spelling?
   a. Jed was disatisfied with the acommodations at his hotel, so he requested another room.
   b. Jed was dissatisfied with the accommodations at his hotel, so he requested another room.
   c. Jed was dissatisfied with the accomodations at his hotel, so he requested another room.
   d. Jed was disatisfied with the accommodations at his hotel, so he requested another room.

8. Which of the following examples is a compound sentence?
   a. Shawn and Jerome played soccer in the backyard for two hours.
   b. Marissa last saw Elena and talked to her this morning.
   c. The baby was sick, so I decided to stay home from work.
   d. Denise, Kurt, and Eric went for a run after dinner.

9. Which of the following examples uses correct punctuation?
   a. Recommended supplies for the hunting trip include the following: rain gear, large backpack, hiking boots, flashlight, and non-perishable foods.
   b. I left the store, because I forgot my wallet.
   c. As soon as the team checked into the hotel; they met in the lobby for a group photo.
   d. None of the furniture came in on time: so they weren't able to move in to the new apartment.

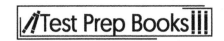

10. Which of the following sentences shows correct word usage?
    a. Your going to have to put you're jacket over their.
    b. You're going to have to put your jacket over there.
    c. Your going to have to put you're jacket over they're.
    d. You're going to have to put your jacket over their.

11. What is the structure of the following sentence?
    The restaurant is unconventional because it serves both Chicago style pizza and New York style pizza.

    a. Simple
    b. Compound
    c. Complex
    d. Compound-complex

12. The following sentence contains what kind of error?
    This summer, I'm planning to travel to Italy, take a Mediterranean cruise, going to Pompeii, and eat a lot of Italian food.

    a. Parallelism
    b. Sentence fragment
    c. Misplaced modifier
    d. Subject-verb agreement

13. The following sentence contains what kind of error?
    Forgetting that he was supposed to meet his girlfriend for dinner, Anita was mad when Fred showed up late.

    a. Parallelism
    b. Run-on sentence
    c. Misplaced modifier
    d. Subject-verb agreement

14. A student writes the following in an essay:
    Protestors filled the streets of the city. Because they were dissatisfied with the government's leadership.

Which of the following is an appropriately-punctuated correction for this sentence?
    a. Protestors filled the streets of the city, because they were dissatisfied with the government's leadership.
    b. Protesters, filled the streets of the city, because they were dissatisfied with the government's leadership.
    c. Because they were dissatisfied with the government's leadership protestors filled the streets of the city.
    d. Protestors filled the streets of the city because they were dissatisfied with the government's leadership.

15. What is the part of speech of the underlined word in the sentence?
    We need to come up with a fresh <u>approach</u> to this problem.

    a. Noun
    b. Verb
    c. Adverb
    d. Adjective

16. What is the noun phrase in the following sentence?
    Charlotte's new German shepherd puppy is energetic.

    a. Puppy
    b. Charlotte
    c. German shepherd puppy
    d. Charlotte's new German shepherd puppy

17. Which word choices will correctly complete the sentence?
    Increasing the price of bus fares has had a greater [affect / effect] on ridership [then / than] expected.

    a. affect; then
    b. affect; than
    c. effect; then
    d. effect; than

18. The following is an example of what type of sentence?
    Although I wished it were summer, I accepted the change of seasons, and I started to appreciate the fall.

    a. Compound
    b. Simple
    c. Complex
    d. Compound-Complex

19. A student reads the following sentence:
    A hundred years ago, automobiles were rare, but now cars are ubiquitous.

However, she doesn't know what the word *ubiquitous* means. Which key context clue is essential to decipher the word's meaning?

    a. Ago
    b. Cars
    c. Now
    d. Rare

20. Which of the examples uses the correct plural form?
    a. Tomatos
    b. Analysis
    c. Cacti
    d. Criterion

Math

Concepts, Data Interpretation, and Problem Solving

1. Which of the following numbers has the greatest value?
    a. 1.4378
    b. 1.07548
    c. 1.43592
    d. 0.89409

2. How will the following number be written in standard form: $(1 \times 10^4) + (3 \times 10^3) + (7 \times 10^1) + (8 \times 10^0)$
    a. 137
    b. 13,078
    c. 1,378
    d. 8,731

3. What is the value of the expression: $7^2 - 3 \times (4 + 2) + 15 \div 5$?
    a. 12.2
    b. 40.2
    c. 34
    d. 58.2

4. Four people split a bill. The first person pays for $\frac{1}{5}$, the second person pays for $\frac{1}{4}$, and the third person pays for $\frac{1}{3}$. What fraction of the bill does the fourth person pay?
    a. $\frac{13}{60}$

    b. $\frac{47}{60}$

    c. $\frac{1}{4}$

    d. $\frac{4}{15}$

5. A student gets an 85% on a test with 20 questions. How many answers did the student solve correctly?
    a. 15
    b. 16
    c. 17
    d. 18

6. If $-3(x + 4) \geq x + 8$, what is the value of x?
    a. $x = 4$
    b. $x \geq 2$
    c. $x \geq -5$
    d. $x \leq -5$

7. What is the value of $x^2 - 2xy + 2y^2$ when $x = 2, y = 3$?
    a. 8
    b. 10
    c. 12
    d. 14

8. If $\sqrt{1 + x} = 4$, what is x?
    a. 10
    b. 15
    c. 20
    d. 25

9. What is the next number in the following series: $1, 3, 6, 10, 15, 21, \dots$?
    a. 26
    b. 27
    c. 28
    d. 29

10. The perimeter of a 6-sided polygon is 56 cm. The length of three sides are 9 cm each. The length of two other sides are 8 cm each. What is the length of the missing side?
    a. 11 cm
    b. 12 cm
    c. 13 cm
    d. 10 cm

11. Katie works at a clothing company and sold 192 shirts over the weekend. $\frac{1}{3}$ of the shirts that were sold were patterned, and the rest were solid. Which mathematical expression would calculate the number of solid shirts Katie sold over the weekend?
    a. $192 \times \frac{1}{3}$
    b. $192 \div \frac{1}{3}$
    c. $192 \times (1 - \frac{1}{3})$
    d. $192 \div 3$

12. Which of the following inequalities is equivalent to $3 - \frac{1}{2}x \geq 2$?
    a. $x \geq 2$
    b. $x \leq 2$
    c. $x \geq 1$
    d. $x \leq 1$

13. If $4x - 3 = 5$, then $x =$
    a. 1
    b. 2
    c. 3
    d. 4

14. Which four-sided shape is always a rectangle?
    a. Rhombus
    b. Square
    c. Parallelogram
    d. Quadrilateral

15. A line passes through the point (1, 2) and crosses the y-axis at $y = 1$. Which of the following is an equation for this line?
    a. $y = 2x$
    b. $y = x + 1$
    c. $x + y = 1$
    d. $y = \frac{x}{2} - 2$

16. A company invests $50,000 in a building where they can produce saws. If the cost of producing one saw is $40, then which function expresses the amount of money the company pays? The variable y is the money paid and x is the number of saws produced.
    a. $y = 50{,}000x + 40$
    b. $y + 40 = x - 50{,}000$
    c. $y = 40x - 50{,}000$
    d. $y = 40x + 50{,}000$

17. A rectangle was formed out of pipe cleaner. Its length was $\frac{1}{2}$ feet and its width was $\frac{11}{2}$ inches. What is its area in square inches?
    a. $\frac{11}{4}$ inch^2
    b. $\frac{11}{2}$ inch^2
    c. 22 inch^2
    d. 33 inch^2

18. Which of the following represent one hundred eighty-two billion, thirty-six thousand, four hundred twenty-one and three hundred fifty-six thousandths?
    a. 182,036,421.356
    b. 182,036,421.0356
    c. 182,000,036,421.0356
    d. 182,000,036,421.356

19. A solution needs 5 mL of saline for every 8 mL of medicine given. How much saline is needed for 45 mL of medicine?
    a. $\frac{225}{8}$ mL
    b. 72 mL
    c. 28 mL
    d. $\frac{45}{8}$ mL

20. What is the 42nd item in the pattern: ▲○○□ ▲○○□ ▲ ...?

    a. ○

    b. ▲

    c. □

    d. None of the above

21. For a group of 20 men, the median weight is 180 pounds and the range is 30 pounds. If each man gains 10 pounds, which of the following would be true?

    a. The median weight will increase, and the range will remain the same.

    b. The median weight and range will both remain the same.

    c. The median weight will stay the same, and the range will increase.

    d. The median weight and range will both increase.

22. Five students take a test. The scores of the first four students are 80, 85, 75, and 60. If the median score is 80, which of the following could NOT be the score of the fifth student?

    a. 60

    b. 80

    c. 85

    d. 100

23. Ten students take a test. Five students get a 50. Four students get a 70. If the average score is 55, what was the last student's score?

    a. 20

    b. 40

    c. 50

    d. 60

24. Given the value of a given stock at monthly intervals, which graph should be used to best represent the trend of the stock?

    a. Box plot

    b. Line plot

    c. Line graph

    d. Circle graph

25. A six-sided die is rolled. What is the probability that the roll is 1 or 2?

    a. $\frac{1}{6}$

    b. $\frac{1}{4}$

    c. $\frac{1}{3}$

    d. $\frac{1}{2}$

26. What is the solution to $9 \times 9 \div 9 + 9 - 9 \div 9$?

    a. 0

    b. 17

    c. 81

    d. 9

27. A grocery store is selling individual bottles of water, and each bottle contains 750 milliliters of water. If 12 bottles are purchased, what conversion will correctly determine how many liters that customer will take home?

    a. 100 milliliters equals 1 liter

    b. 1,000 milliliters equals 1 liter

    c. 1,000 liters equals 1 milliliter

    d. 10 liters equals 1 milliliter

28. Which of the following statements is true about the two lines below?

    a. The two lines are parallel but not perpendicular.

    b. The two lines are perpendicular but not parallel.

    c. The two lines are both parallel and perpendicular.

    d. The two lines are neither parallel nor perpendicular.

29. Which common denominator would be used to evaluate $\frac{2}{3} + \frac{4}{5}$?

    a. 15

    b. 3

    c. 5

    d. 10

30. An equilateral triangle has a perimeter of 18 feet. If a square whose sides have the same length as one side of the triangle is built, what will be the area of the square?

    a. 6 square feet

    b. 36 square feet

    c. 256 square feet

    d. 1000 square feet

31. The area of a given rectangle is 24 centimeters. If the measure of each side is multiplied by 3, what is the area of the new figure?

    a. 48 cm^2

    b. 72 cm^2

    c. 216 cm^2

    d. 13,824 cm^2

32. Apples cost $2 each, while bananas cost $3 each. Maria purchased 10 fruits in total and spent $22. How many apples did she buy?

    a. 5

    b. 6

    c. 7

    d. 8

Estimation

Directions: Estimate the answer in your head. No writing is permitted. An exact answer is not expected.

33. Estimate the sum of 3.3 + 2.15 + 4.
    a. 6
    b. 8
    c. 9
    d. 11

34. Estimate the decimal equivalent of $\frac{3}{25}$.
    a. 0.25
    b. 0.4
    c. 0.9
    d. 0.12

35. 6 is about 30% of what number?
    a. 16
    b. 20
    c. 24
    d. 26

36. What is the approximate value of the following expression?
$$\sqrt{8.2^2 + 5.75^2}$$
    a. 14
    b. 10
    c. 9
    d. 100

37. Estimate the solution to 800 ÷ 9.
    a. 80
    b. 10
    c. 400
    d. 20

38. Estimate the product of 5.88 × 3.
    a. 18
    b. 15
    c. 20
    d. 12

39. If Danny takes 35 minutes to run 3 miles, about how long should it take him to run 5 miles maintaining the same speed?
    a. 80 min
    b. 50 min
    c. 60 min
    d. 90 min

40. Last year, the New York City area received approximately $27\frac{3}{4}$ inches of snow. The Denver area received approximately 3 times as much snow as New York City. About how much snow fell in Denver?

    a. 60 inches
    b. 96 inches
    c. 9 inches
    d. 84 inches

41. This chart indicates how many sales of CDs, vinyl records, and MP3 downloads occurred over the last year. Approximately what percentage of the total sales was from CDs?

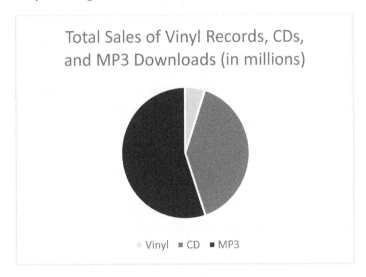

    a. 55%
    b. 25%
    c. 40%
    d. 5%

42. After a 20% sale discount, Frank purchased a new refrigerator for $810. About how much did he save from the original price?

    a. $150
    b. $200
    c. $100
    d. $250

43. A school has 14 teachers and 20 teaching assistants. They have 200 students. What is the approximate ratio of faculty to students?

    a. 3:2
    b. 2:3
    c. 6:1
    d. 1:6

44. A student gets a 90% on a test with 20 questions. About how many answers did the student solve correctly?
    a. 20
    b. 16
    c. 14
    d. 18

45. If Sarah reads at an average rate of 21 pages in four nights, about how long will it take her to read 102 pages?
    a. 10 nights
    b. 25 nights
    c. 8 nights
    d. 20 nights

46. Alan currently weighs 200 pounds, but he wants to lose weight to get down to 175 pounds. What is this approximate difference in kilograms? (1 pound is approximately equal to 0.45 kilograms.)
    a. 9 kg
    b. 12.5 kg
    c. 25 kg
    d. 5 kg

47. Johnny earns $2434.50 from his job each month. He pays $1437 for monthly expenses. About how much will Johnny have left over from three months' of saving?
    a. $2500
    b. $1000
    c. $2000
    d. $3000

48. What is $\frac{420}{98}$ rounded to the nearest integer?
    a. 3
    b. 4
    c. 5
    d. 6

49. What is the estimated solution to the following expression?
$$4.2 \times 7 + (25 - 21)^2 \div 2$$
    a. 512
    b. 36
    c. 60.5
    d. 22

50. In Jim's school, there are 3 girls for every 2 boys. There are 600 students in total. Using this information, about how many students are girls?
    a. 260
    b. 120
    c. 60
    d. 360

Ability

Figure Matrices

For the section below, select the answer that best completes the puzzle. Note that each of the figures in these matrices follow a certain rule. One rule applies to the rows, and another rule applies to the columns.

1.

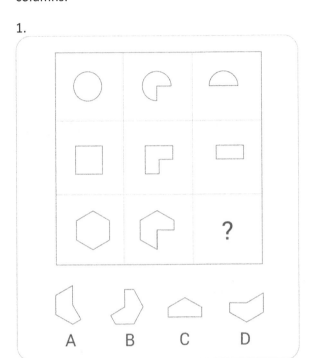

2.

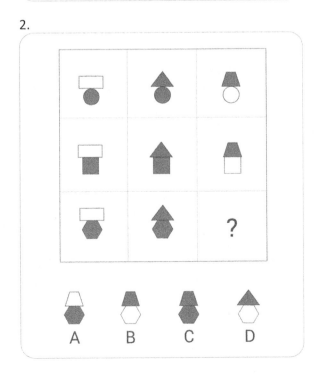

3.

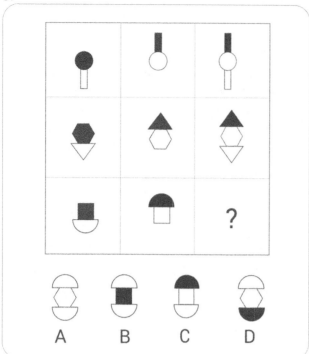

4.

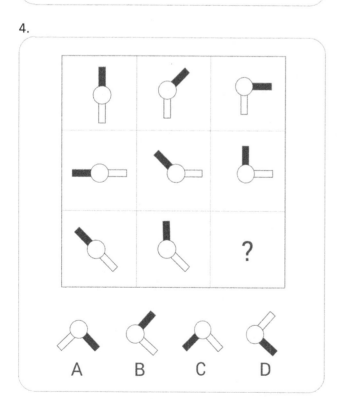

5.

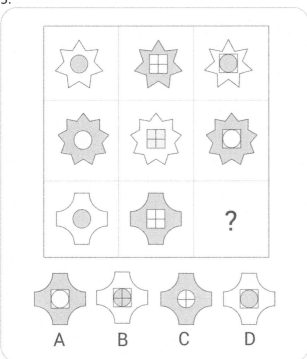

6.

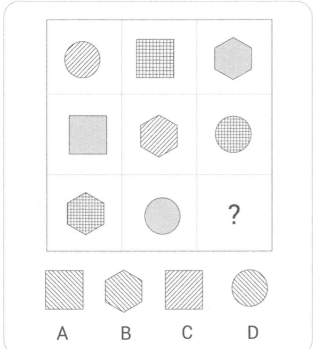

7.

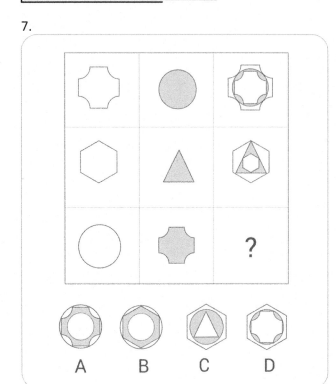

8.

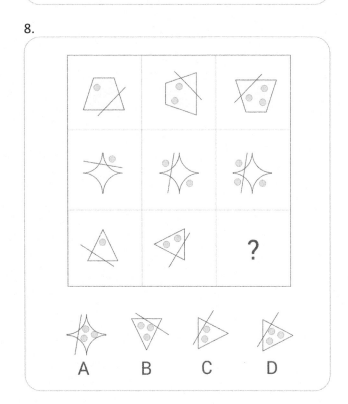

9.

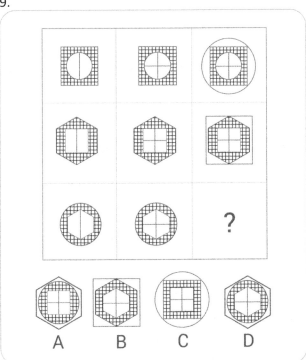

10.

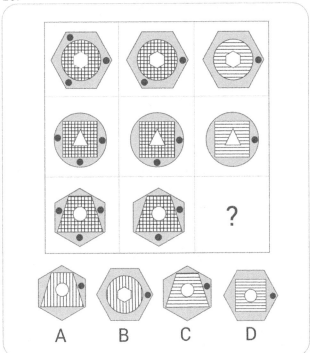

11.

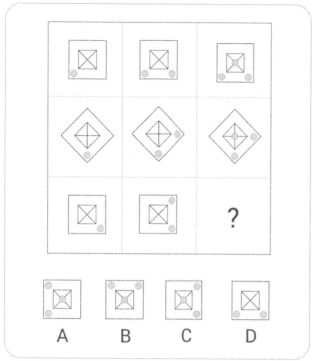

12.

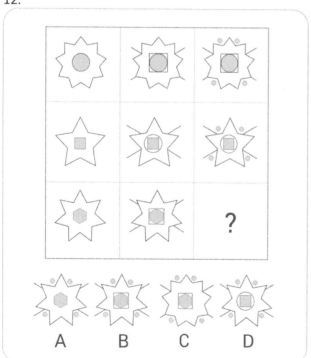

13.

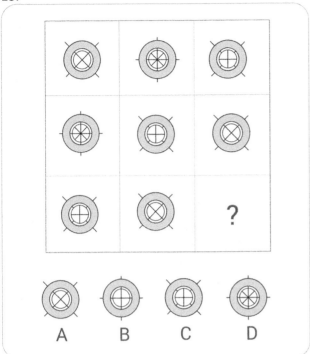

14.

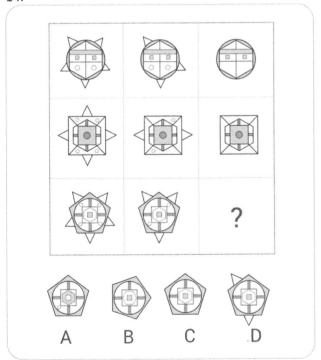

15.

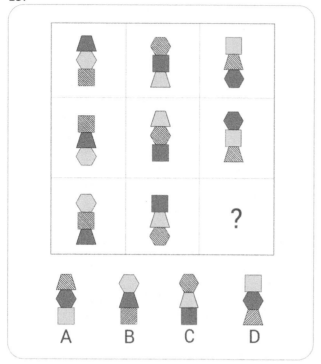

16.

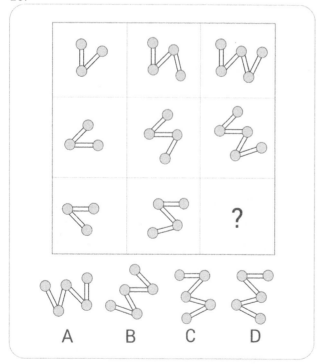

17.

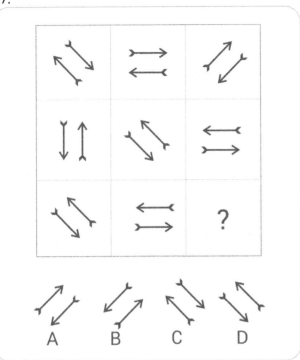

18.

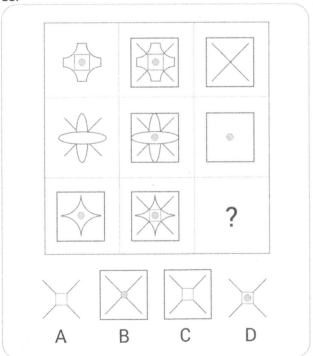

19.

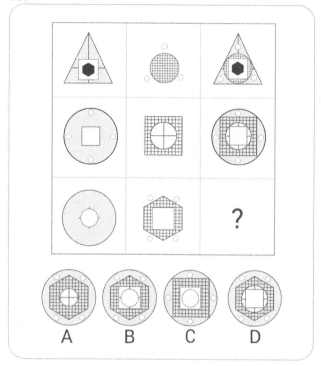

20.

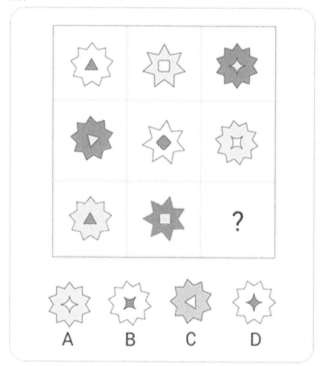

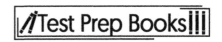

Paper Folding

For this section, imagine each shaded square is a piece of paper that is being folded. The arrow tells you which way to fold the paper. Once the paper is folded, holes are punched into the paper. Once you "reopen" the paper with your imagination, pick the answer choice that represents the unfolded piece of paper.

1.

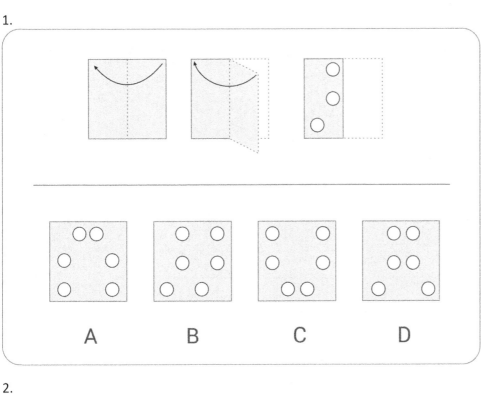

2.

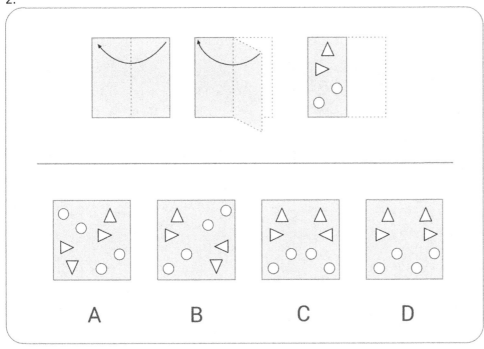

3.

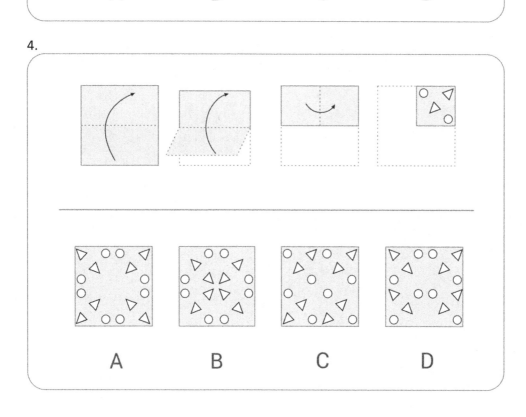

4.

5.

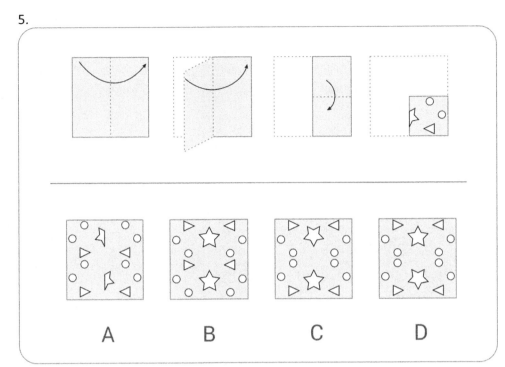

A   B   C   D

6.

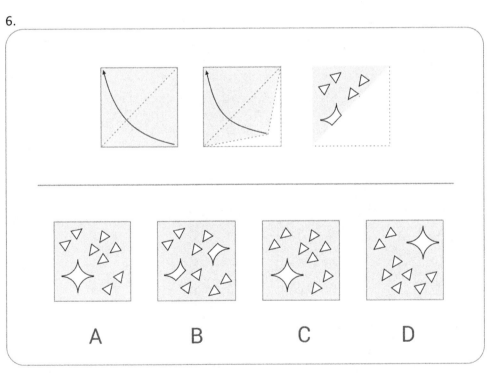

A   B   C   D

7.

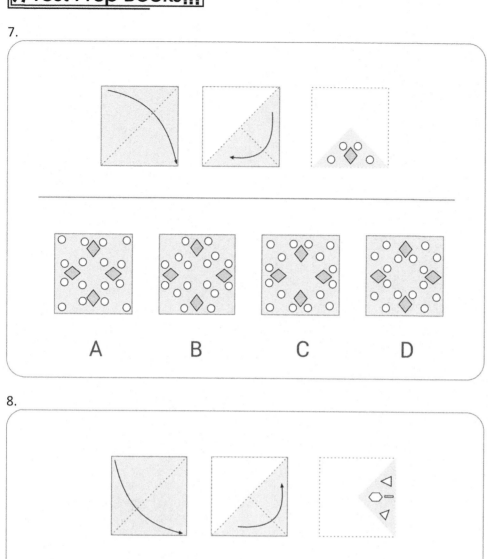

8.

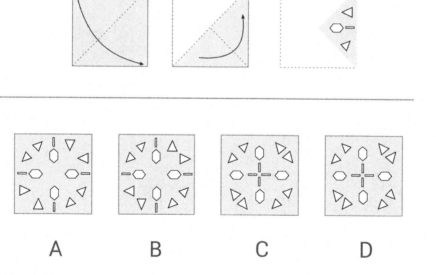

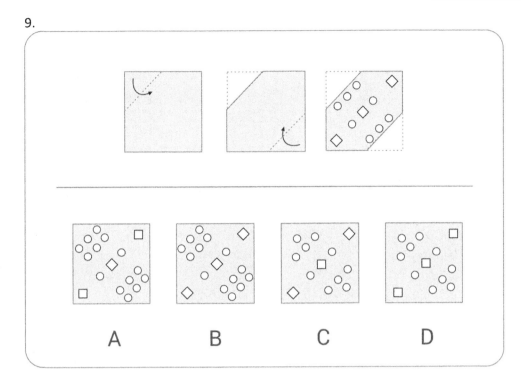

9.

10.

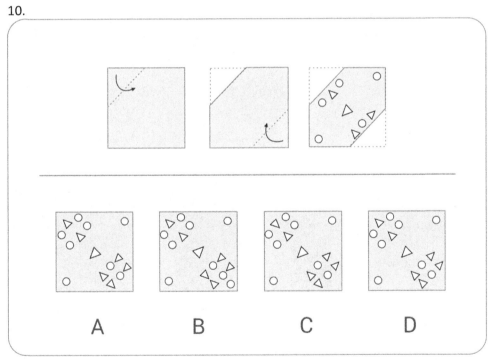

11.

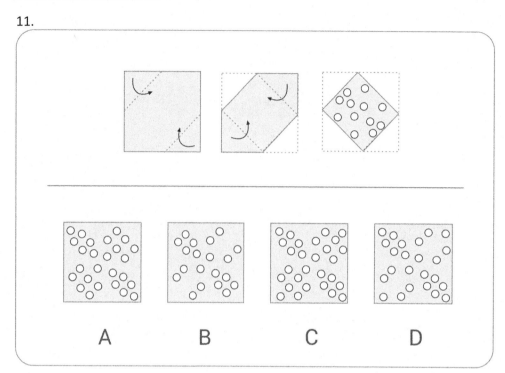

12.

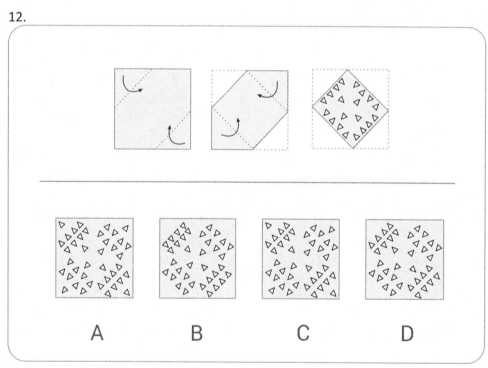

13.

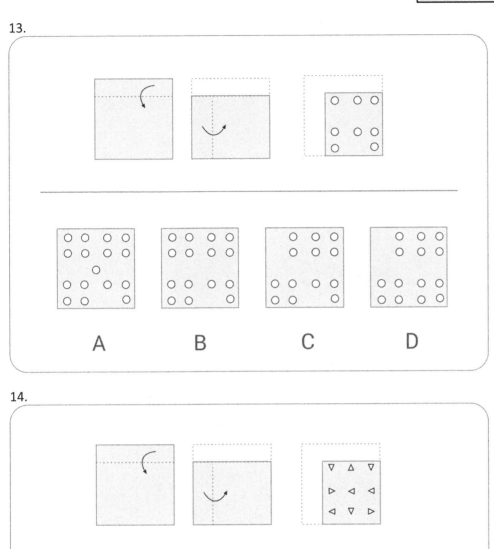

14.

15.

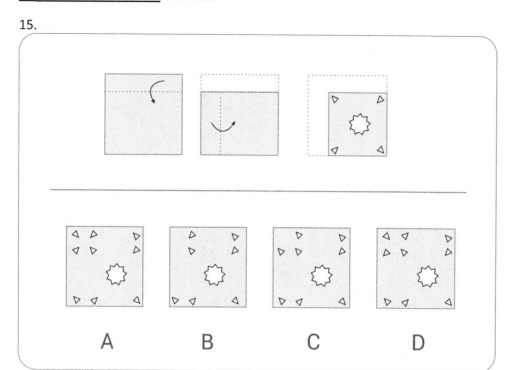

16.

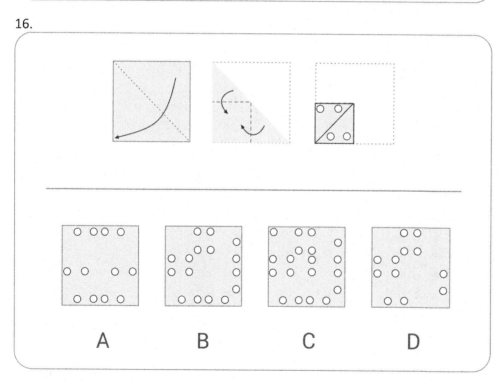

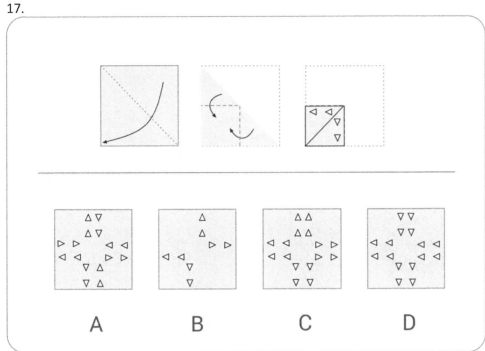

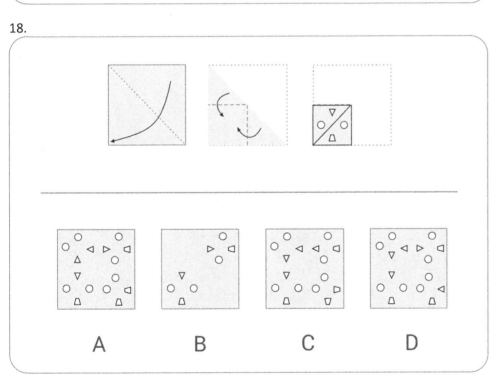

17.

18.

19.

A   B   C   D

20.

A   B   C   D

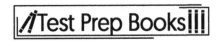

Figure Classifications

In each question, the first three figures have similarities. Choose the answer that goes best with the first three figures.

1.

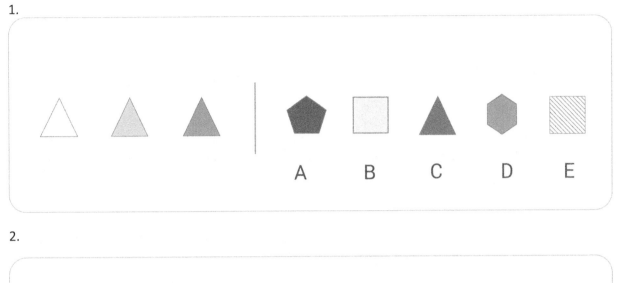

2.

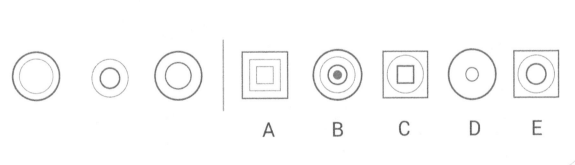

3.

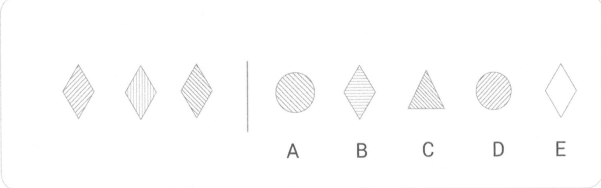

4.

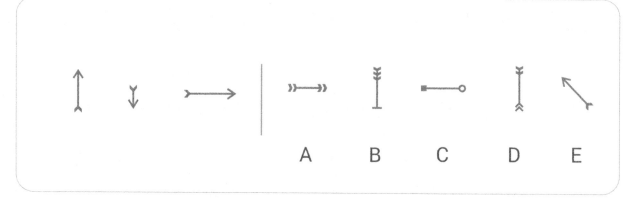

A B C D E

5.

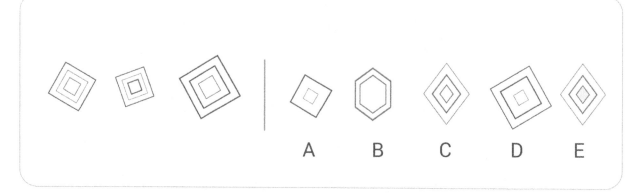

A B C D E

6.

A B C D E

7.

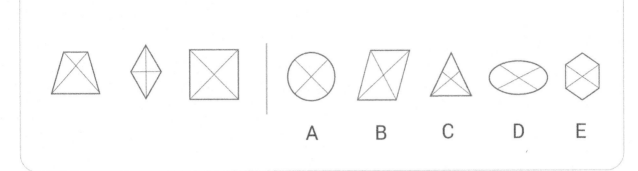

8.

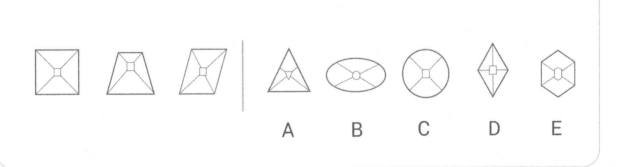

9.

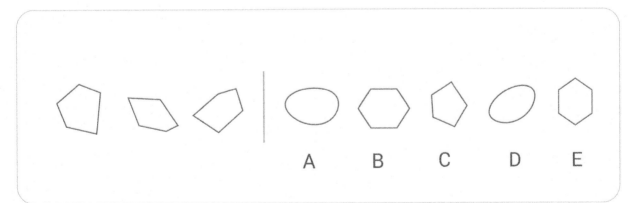

10.

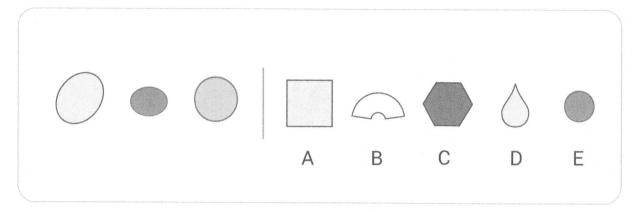

11.

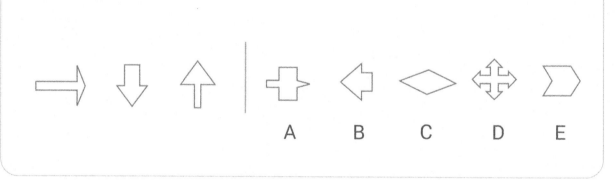

12.

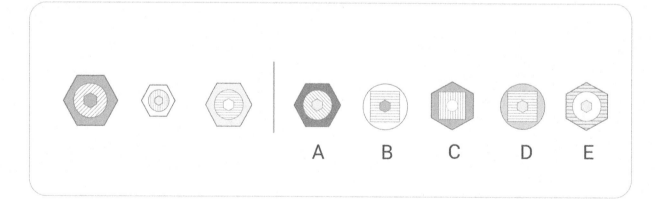

13.

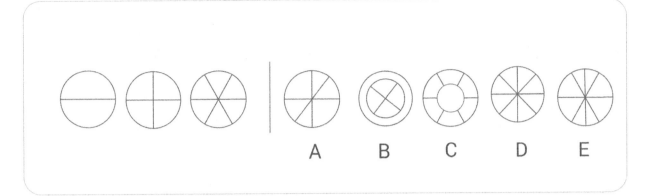

14.

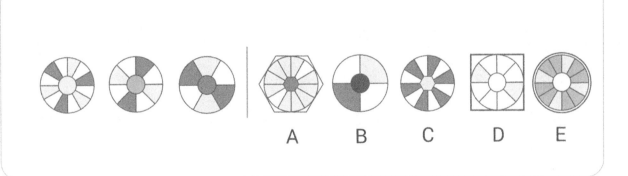

15.

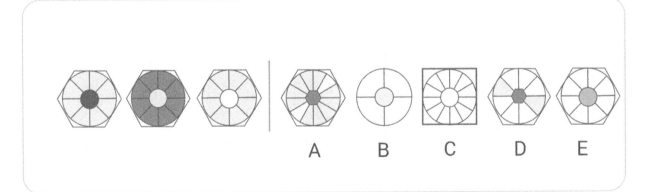

16.

17.

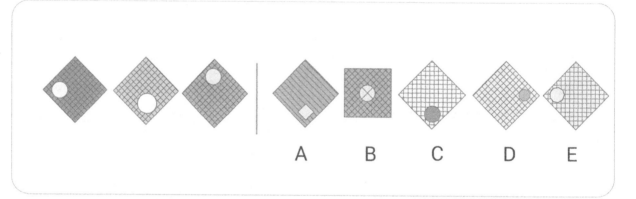

18.

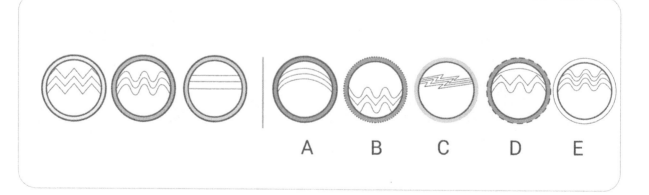

19.

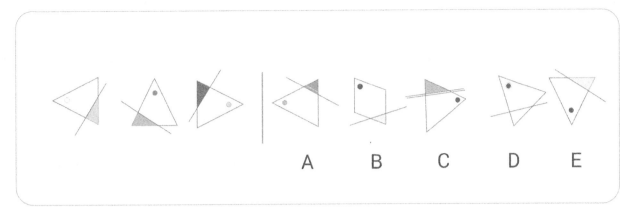

20.

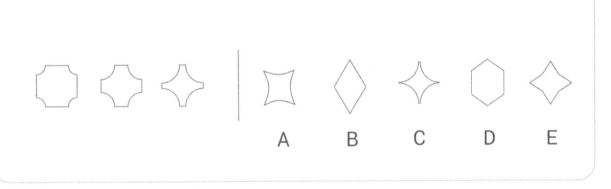

21.

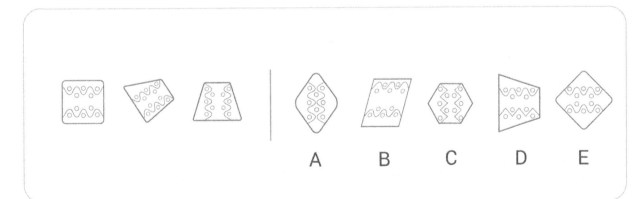

22.

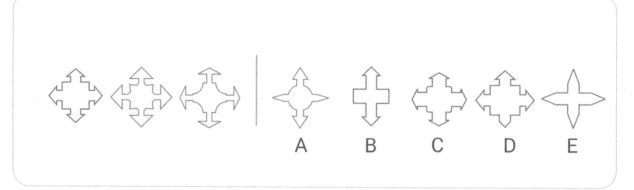

23.

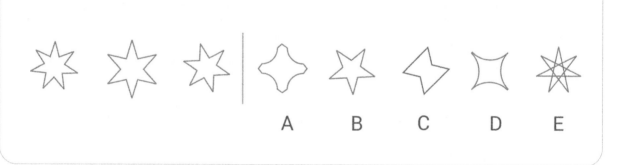

Answer Explanations

Reading

1. C: Gulliver becomes acquainted with the people and practices of his new surroundings. Choice *C* is the correct answer because it most extensively summarizes the entire passage. While Choices *A* and *B* are reasonable possibilities, they reference portions of Gulliver's experiences, not the whole. Choice *D* is incorrect because Gulliver doesn't express repentance or sorrow in this particular passage.

2. A: Principal refers to *chief* or *primary* within the context of this text. Choice *A* is the answer that most closely aligns with this answer. Choices *B* and *D* make reference to a helper or followers while Choice *C* doesn't meet the description of Gulliver from the passage.

3. C: One can reasonably infer that Gulliver is considerably larger than the children who were playing around him because multiple children could fit into his hand. Choice *B* is incorrect because there is no indication of stress in Gulliver's tone. Choices *A* and *D* aren't the best answer because though Gulliver seems fond of his new acquaintances, he didn't travel there with the intentions of meeting new people or to express a definite love for them in this particular portion of the text.

4. C: The emperor made a *definitive decision* to expose Gulliver to their native customs. In this instance, the word *mind* was not related to a vote, question, or cognitive ability.

5. A: Choice *A* is correct. This assertion does *not* support the fact that games are a commonplace event in this culture because it mentions conduct, not games. Choices *B*, *C*, and *D* are incorrect because these do support the fact that games were a commonplace event.

6. B: Choice *B* is the only option that mentions the correlation between physical ability and leadership positions. Choices *A* and *D* are unrelated to physical strength and leadership abilities. Choice *C* does not make a deduction that would lead to the correct answer—it only comments upon the abilities of common townspeople.

7. D: It emphasizes Mr. Utterson's anguish in failing to identify Hyde's whereabouts. Context clues indicate that Choice *D* is correct because the passage provides great detail of Mr. Utterson's feelings about locating Hyde. Choice *A* does not fit because there is no mention of Mr. Lanyon's mental state. Choice *B* is incorrect; although the text does make mention of bells, Choice *B* is not the *best* answer overall. Choice *C* is incorrect because the passage clearly states that Mr. Utterson was determined, not unsure.

8. A: In the city. The word *city* appears in the passage several times, thus establishing the location for the reader.

9. B: It scares children. The passage states that the Juggernaut causes the children to scream. Choices *A* and *D* don't apply because the text doesn't mention either of these instances specifically. Choice *C* is incorrect because there is nothing in the text that mentions space travel.

10. B: To constantly visit. The mention of *morning, noon,* and *night* make it clear that the word *haunt* refers to frequent appearances at various locations. Choice *A* doesn't work because the text makes no mention of levitating. Choices *C* and *D* are not correct because the text makes mention of Mr. Utterson's anguish and disheartenment because of his failure to find Hyde but does not make mention of Mr. Utterson's feelings negatively affecting anyone else.

11. D: This is an example of alliteration. Choice *D* is the correct answer because of the repetition of the *L*-words. Hyperbole is an exaggeration, so Choice *A* doesn't work. No comparison is being made, so no simile or metaphor is being used, thus eliminating Choices *B* and *C*.

12. D: The speaker intends to continue to look for Hyde. Choices *A* and *B* are not possible answers because the text doesn't refer to any name changes or an identity crisis, despite Mr. Utterson's extreme obsession with finding Hyde. The text also makes no mention of a mistaken identity when referring to Hyde, so Choice *C* is also incorrect.

13. A: The tone is exasperated. While contemplative is an option because of the inquisitive nature of the text, Choice *A* is correct because the speaker is annoyed by the thought of being included when he felt that the fellow members of his race were being excluded. The speaker is not nonchalant, nor accepting of the circumstances which he describes.

14. C: Choice *C*, *contented*, is the only word that has different meaning. Furthermore, the speaker expresses objection and disdain throughout the entire text.

15. B: To address the feelings of exclusion expressed by African Americans after the establishment of the Fourth of July holiday. While the speaker makes biblical references, it is not the main focus of the passage, thus eliminating Choice *A* as an answer. The passage also makes no mention of wealthy landowners and doesn't speak of any positive response to the historical events, so Choices *C* and *D* are not correct.

16. D: Choice *D* is the correct answer because it clearly makes reference to justice being denied.

17. D: Hyperbole. Choices *A* and *B* are unrelated. Assonance is the repetition of sounds and commonly occurs in poetry. Parallelism refers to two statements that correlate in some manner. Choice *C* is incorrect because amplification normally refers to clarification of meaning by broadening the sentence structure, while hyperbole refers to a phrase or statement that is being exaggerated.

18. C: Choice *C* is correct because the speaker is clear about his intention and stance throughout the text; thus, it's not true that he makes biblical references to display his own equivocation and that of those that he represents. Choice *A* could be true, but the words "common text" is arguable because not everyone will understand the reference. Choice *B* is also partially true, as another group of people affected by slavery are being referenced. However, the speaker is not trying to convince the audience that injustices have been committed, as it is already understood there have been injustices committed. Choice *D* is also close to the correct answer, but it is not the best answer choice possible.

19. B: A period of time. It is apparent that Lincoln is referring to a period of time within the context of the passage because of how the sentence is structured with the word *ago*.

20. C: Lincoln's reference to *the brave men, living and dead, who struggled here,* proves that he is referring to a battlefield. Choices *A* and *B* are incorrect, as a *civil war* is mentioned and not a war with France or a war in the Sahara Desert. Choice *D* is incorrect because it does not make sense to consecrate a President's ground instead of a battlefield ground for soldiers who died during the American Civil War.

21. D: Abraham Lincoln is the former president of the United States, and he references a "civil war" during his address.

22. A: The audience should consider the death of the people that fought in the war as an example and perpetuate the ideals of freedom that the soldiers died fighting for. Lincoln doesn't address any of the topics outlined in Choices *B*, *C*, or *D*. Therefore, Choice *A* is the correct answer.

23. D: Choice *D* is the correct answer because of the repetition of the word *people* at the end of the passage. Choice *A*, *antimetabol*, is the repetition of words in a succession. Choice *B*, *antiphrasis*, is a form of denial of an assertion in a text. Choice *C*, *anaphora*, is the repetition that occurs at the beginning of sentences.

24. A: Choice *A* is correct because Lincoln's intention was to memorialize the soldiers who had fallen as a result of war as well as celebrate those who had put their lives in danger for the sake of their country. Choices *B* and *D* are incorrect because Lincoln's speech was supposed to foster a sense of pride among the members of the audience while connecting them to the soldiers' experiences.

25. A: The word *patronage* most nearly means *auspices*, which means *protection* or *support*. Choice *B*, *aberration*, means *deformity* and does not make sense within the context of the sentence. Choice *C*, *acerbic,* means *bitter* and also does not make sense in the sentence. Choice *D*, *adulation*, is a positive word meaning *praise*, and thus does not fit with the word *condescending* in the sentence.

26. D: *Working man* is most closely aligned with Choice *D*, *bourgeois.* In the context of the speech, the word *bourgeois* means *working* or *middle class*. Choice *A*, *Plebeian*, does suggest *common people*; however, this is a term that is specific to ancient Rome. Choice *B*, *viscount*, is a European title used to describe a specific degree of nobility. Choice *C*, *entrepreneur*, is a person who operates their own business.

27. C: In the context of the speech, the term *working man* most closely correlates with Choice *C*, *working man is someone who works for wages among the middle class*. Choice *A* is not mentioned in the passage and is off-topic. Choice *B* may be true in some cases, but it does not reflect the sentiment described for the term *working man* in the passage. Choice *D* may also be arguably true. However, it is not given as a definition but as *acts* of the working man, and the topics of *field, factory,* and *screen* are not mentioned in the passage.

28. D: *Enterprise* most closely means *cause*. Choices *A, B,* and *C* are all related to the term *enterprise*. However, Dickens speaks of a *cause* here, not a company, courage, or a game. *He will stand by such an enterprise* is a call to stand by a cause to enable the working man to have a certain autonomy over his own economic standing. The very first paragraph ends with the statement that the working man *shall . . . have a share in the management of an institution which is designed for his benefit.*

29. B: The speaker's salutation is one from an entertainer to his audience, and uses the friendly language to connect to his audience before a serious speech. Recall in the first paragraph that the speaker is there to "accompany [the audience] . . . through one of my little Christmas books," making him an author there to entertain the crowd with his own writing. The speech preceding the reading is the passage itself, and, as the tone indicates, a serious speech addressing the "working man." Although the passage speaks of employers and employees, the speaker himself is not an employer of the audience, so Choice *A* is incorrect. Choice *C* is also incorrect, as the salutation is not used ironically, but sincerely, as the speech addresses the wellbeing of the crowd. Choice *D* is incorrect because the speech is not given by a politician, but by a writer.

30. B: For the working man to have a say in his institution which is designed for his benefit Choice *A* is incorrect because that is the speaker's *first* desire, not his second. Choices *C* and *D* are tricky because the language of both of these is mentioned after the word *second*. However, the speaker doesn't get to the second wish until the next sentence. Choices *C* and *D* are merely prepositions preparing for the statement of the main clause, Choice *B*.

Written Expression

1. B: In Choice B the word *Uncle* should not be capitalized, because it is not functioning as a proper noun. If the word named a specific uncle, such as *Uncle Jerry*, then it would be considered a proper noun and should be capitalized. Choice *A* correctly capitalizes the proper noun *East Coast*, and does not capitalize *winter*, which functions as a common noun in the sentence. Choice *C* correctly capitalizes the name of a specific college course, which is considered a proper noun. Choice *D* correctly capitalizes the proper noun *Jersey Shore*.

2. A: Choice *A* is correctly punctuated because it uses a semicolon to join two independent clauses that are related in meaning. Each of these clauses could function as an independent sentence. Choice *B* is incorrect because the conjunction is not preceded by a comma. A comma and conjunction should be used together to join independent clauses. Choice *C* is incorrect because a comma should only be used to join independent sentences when it also includes a coordinating conjunction such as *and* or *so*. Choice *D* does not use punctuation to join the independent clauses, so it is considered a fused (same as a run-on) sentence.

3. C: *All of Shannon's family and friends* is the complete subject because it includes who or what is doing the action in the sentence as well as the modifiers that go with it. Choice *A* is incorrect because it only includes the simple subject of the sentence. Choices B and D are incorrect because they only include part of the complete subject.

4. D: Choice *D* correctly places a hyphen after the prefix *ex* to join it to the word *husband*. Words that begin with the prefixes *great*, *trans*, *ex*, *all*, and *self*, require a hyphen. Choices A and C place hyphens in words where they are not needed. *Beagle mix* would only require a hyphen if coming before the word *Henry*, since it would be serving as a compound adjective in that instance. Choice *B* contains hyphens that are in the correct place but are formatted incorrectly since they include spaces between the hyphens and the surrounding words.

5. C: Choice *C* is correct because quotation marks should be used for the title of a short work such as a poem. Choices A, B, and D are incorrect because the titles of novels, films, and newspapers should be placed in italics, not quotation marks.

6. C: This question focuses on the correct usage of the commonly confused word pairs of *it's/its* and *then/than*. *It's* is a contraction for *it is* or *it has*. *Its* is a possessive pronoun. The word *than* shows comparison between two things. *Then* is an adverb that conveys time. Choice *C* correctly uses *it's* and *than*. *It's* is a contraction for *it has* in this sentence, and *than* shows comparison between *work* and *rest*. None of the other answer choices use both of the correct words.

7. B: *Dissatisfied* and *accommodations* are both spelled correctly in Choice *B*. These are both considered commonly misspelled words. One or both words are spelled incorrectly in choices A, C, and D.

8. C: Choice *C* is a compound sentence because it joins two independent clauses—*The baby was sick* and *I decided to stay home from work*—with a comma and the coordinating conjunction *so*. Choices A, B, and D, are all simple sentences, each containing one independent clause with a complete subject and predicate. Choices A and D each contain a compound subject, or more than one subject, but they are

still simple sentences that only contain one independent clause. Choice *B* contains a compound verb (more than one verb), but it's still a simple sentence.

9. A: In this example, a colon is correctly used to introduce a series of items. Choice *B* places an unnecessary comma before the word *because*. A comma is not needed before the word *because* when it introduces a dependent clause at the end of a sentence and provides necessary information to understand the sentence. Choice *C* is incorrect because it uses a semi-colon instead of a comma to join a dependent clause and an independent clause. Choice *D* is incorrect because it uses a colon in place of a comma and coordinating conjunction to join two independent clauses.

10. B: Choice *B* correctly uses the contraction for *you are* as the subject of the sentence, and it correctly uses the possessive pronoun *your* to indicate ownership of the jacket. It also correctly uses the adverb *there*, indicating place. Choice *A* is incorrect because it reverses the possessive pronoun *your* and the contraction for *you are*. It also uses the possessive pronoun *their* instead of the adverb *there*. Choice *C* is incorrect because it reverses *your* and *you're* and uses the contraction for *they are* in place of the adverb *there*. Choice *D* incorrectly uses the possessive pronoun *their* instead of the adverb *there*.

11. C: A complex sentence joins an independent or main clause with a dependent or subordinate clause. In this case, the main clause is "The restaurant is unconventional." This is a clause with one subject-verb combination that can stand alone as a grammatically-complete sentence. The dependent clause is "because it serves both Chicago style pizza and New York style pizza." This clause begins with the subordinating conjunction *because* and also consists of only one subject-verb combination. *A* is incorrect because a simple sentence consists of only one verb-subject combination—one independent clause. *B* is incorrect because a compound sentence contains two independent clauses connected by a conjunction. *D* is incorrect because a complex-compound sentence consists of two or more independent clauses and one or more dependent clauses.

12. A: Parallelism refers to consistent use of sentence structure or word form. In this case, the list within the sentence does not utilize parallelism; three of the verbs appear in their base form—*travel*, *take*, and *eat*—but one appears as a gerund—*going*. A parallel version of this sentence would be "This summer, I'm planning to travel to Italy, take a Mediterranean cruise, go to Pompeii, and eat a lot of Italian food." *B* is incorrect because this description is a complete sentence. *C* is incorrect as a misplaced modifier is a modifier that is not located appropriately in relation to the word or words they modify. *D* is incorrect because subject-verb agreement refers to the appropriate conjugation of a verb in relation to its subject.

13. C: In this sentence, the modifier is the phrase "Forgetting that he was supposed to meet his girlfriend for dinner." This phrase offers information about Fred's actions, but the noun that immediately follows it is Anita, creating some confusion about the "do-er" of the phrase. A more appropriate sentence arrangement would be "Forgetting that he was supposed to meet his girlfriend for dinner, Fred made Anita mad when he showed up late." *A* is incorrect as parallelism refers to the consistent use of sentence structure and verb tense, and this sentence is appropriately consistent. *B* is incorrect as a run-on sentence does not contain appropriate punctuation for the number of independent clauses presented, which is not true of this description. *D* is incorrect because subject-verb agreement refers to the appropriate conjugation of a verb relative to the subject, and all verbs have been properly conjugated.

14. D: The problem in the original passage is that the second sentence is a dependent clause that cannot stand alone as a sentence; it must be attached to the main clause found in the first sentence. Because the main clause comes first, it does not need to be separated by a comma. However, if the dependent clause came first, then a comma would be necessary, which is why Choice *C* is incorrect. *A* and *B* also insert unnecessary commas into the sentence.

15. A: A noun refers to a person, place, thing, or idea. Although the word *approach* can also be used as a verb, in the sentence it functions as a noun within the noun phrase "a fresh approach," so *B* is incorrect. An adverb is a word or phrase that provides additional information of the verb, but because the verb is *need* and not *approach*, then *C* is false. An adjective is a word that describes a noun, used here as the word *fresh*, but it is not the noun itself. Thus, *D* is also incorrect.

16. D: A noun phrase consists of the noun and all of its modifiers. In this case, the subject of the sentence is the noun *puppy*, but it is preceded by several modifiers—adjectives that give more information about what kind of puppy, which are also part of the noun phrase. Thus, *A* is incorrect. Charlotte is the owner of the puppy and a modifier of the puppy, so *B* is false. *C* is incorrect because it contains some, but not all, of the modifiers pertaining to the puppy. *D* is correct because it contains all of them.

17. D: In this sentence, the first answer choice requires a noun meaning *impact* or *influence*, so *effect* is the correct answer. For the second answer choice, the sentence is drawing a comparison. *Than* shows a comparative relationship whereas *then* shows sequence or consequence. *A* and *C* can be eliminated because they contain the choice *then*. *B* is incorrect because *affect* is a verb while this sentence requires a noun.

18. D: Since the sentence contains two independent clauses and a dependent clause, the sentence is categorized as compound-complex:

> Independent clause: *I accepted the change of seasons*

> Independent clause: *I started to appreciate the fall*

> Dependent clause: *Although I wished it were summer*

19. D: Students can use context clues to make a careful guess about the meaning of unfamiliar words. Although all of the words in a sentence can help contribute to the overall sentence, in this case, the adjective that pairs with *ubiquitous* gives the most important hint to the student—cars were first *rare*, but now they are *ubiquitous*. The inversion of *rare* is what gives meaning to the rest of the sentence and *ubiquitous* means "existing everywhere" or "not rare." *A* is incorrect because *ago* only indicates a time frame. *B* is incorrect because *cars* does not indicate a contrasting relationship to the word *ubiquitous* to provide a good context clue. *C* is incorrect because it also only indicates a time frame, but used together with *rare*, it provides the contrasting relationship needed to identify the meaning of the unknown word.

20. C: Cacti is the correct plural form of the word *cactus*. Choice *A* (*tomatos*) includes an incorrect spelling of the plural of *tomato*. Both Choice *B* (*analysis*) and Choice *D* (*criterion*) are incorrect because they are in singular form. The correct plural form for these choices would be *criteria* and analyses.

Math

Concepts, Data Interpretation, and Problem Solving

1. A: Compare each numeral after the decimal point to figure out which overall number is greatest. In answers *A* (1.43785) and *C* (1.43592), both have the same tenths (4) and hundredths (3). However, the thousandths is greater in answer *A* (7), so *A* has the greatest value overall.

2. B: 13,078. The power of 10 by which a digit is multiplied corresponds with the number of zeros following the digit when expressing its value in standard form. Therefore:

$$(1 \times 10^4) + (3 \times 10^3) + (7 \times 10^1) + (8 \times 10^0)$$

$$10,000 + 3,000 + 70 + 8$$

$$13,078$$

3. C: 34. When performing calculations consisting of more than one operation, the order of operations should be followed: *Parenthesis, Exponents, Multiplication/Division, Addition/Subtraction*. Parenthesis:

$$7^2 - 3 \times (4 + 2) + 15 \div 5$$

$$7^2 - 3 \times (6) + 15 \div 5$$

Exponents:

$$7^2 - 3 \times 6 + 15 \div 5$$
$$49 - 3 \times 6 + 15 \div 5$$

Multiplication/Division (from left to right):

$$49 - 3 \times 6 + 15 \div 5$$

$$49 - 18 + 3$$

Addition/Subtraction (from left to right):

$$49 - 18 + 3 = 34$$

4. A: To find the fraction of the bill that the first three people pay, the fractions need to be added, which means finding common denominator. The common denominator will be 60.

$$\frac{1}{5} + \frac{1}{4} + \frac{1}{3}$$

$$\frac{12}{60} + \frac{15}{60} + \frac{20}{60}$$

$$\frac{47}{60}$$

The remainder of the bill is:

$$1 - \frac{47}{60} = \frac{60}{60} - \frac{47}{60} = \frac{13}{60}$$

5. C: 85% of a number means multiplying that number by 0.85. So, $0.85 \times 20 = \frac{85}{100} \times \frac{20}{1}$, which can be simplified to:

$$\frac{17}{20} \times \frac{20}{1} = 17$$

6. D: $x \leq -5$. When solving a linear equation or inequality:

Distribution is performed if necessary:

$$-3(x + 4) \rightarrow -3x - 12 \geq x + 8$$

This means that any like terms on the same side of the equation/inequality are combined.

The equation/inequality is manipulated to get the variable on one side. In this case, subtracting x from both sides produces:

$$-4x - 12 \geq 8$$

The variable is isolated using inverse operations to undo addition/subtraction. Adding 12 to both sides produces $-4x \geq 20$.

The variable is isolated using inverse operations to undo multiplication/division. Remember if dividing by a negative number, the relationship of the inequality reverses, so the sign is flipped. In this case, dividing by -4 on both sides produces $x \leq -5$.

7. B: Start with the original equation: x- 2xy + 2y, then replace each instance of x with a 2, and each instance of y with a 3 to get

$$2^2 - 2 \cdot 2 \cdot 3 + 2 \cdot 3^2$$

$$4 - 12 + 18 = 10$$

8. B: Start by squaring both sides to get $1 + x = 16$. Then subtract 1 from both sides to get $x = 15$.

9. C: Each number in the sequence is adding one more than the difference between the previous two. For example:

$$10 - 6 = 4, 4 + 1 = 5$$

Therefore, the next number after 10 is $10 + 5 = 15$.

Going forward, $21 - 15 = 6, 6 + 1 = 7$. The next number is $21 + 7 = 28$. Therefore, the difference between numbers is the set of whole numbers starting at 2: 2, 3, 4, 5, 6, 7....

10. C: The perimeter is found by calculating the sum of all sides of the polygon.

$$9 + 9 + 9 + 8 + 8 + s = 56$$
where s is the missing side length.

Therefore, 43 plus the missing side length is equal to 56. The missing side length is 13 cm.

11. C: $\frac{1}{3}$ of the shirts sold were patterned. Therefore, $1 - \frac{1}{3} = \frac{2}{3}$ of the shirts sold were solid. Anytime "of" a quantity appears in a word problem, multiplication needs to be used. Therefore:

$$192 \times \frac{2}{3} = \frac{192*2}{3} = \frac{384}{3}$$

128 solid shirts were sold

The entire expression is $192 \times \left(1 - \frac{1}{3}\right)$.

12. B: To simplify this inequality, subtract 3 from both sides to get:

$$-\frac{1}{2}x \geq -1$$

Then, multiply both sides by -2 (remembering this flips the direction of the inequality) to get $x \leq 2$.

13. B: Add 3 to both sides to get $4x = 8$. Then divide both sides by 4 to get $x = 2$.

14. B: A rectangle is a specific type of parallelogram. It has 4 right angles. A square is a rhombus that has 4 right angles. Therefore, a square is always a rectangle because it has two sets of parallel lines and 4 right angles.

15. B: From the slope-intercept form, $y = mx + b$, it is known that b is the y-intercept, which is 1. Compute the slope as $\frac{2-1}{1-0} = 1$, so the equation should be $y = x + 1$.

16. D: For manufacturing costs, there is a linear relationship between the cost to the company and the number produced, with a y-intercept given by the base cost of acquiring the means of production, and a slope given by the cost to produce one unit. In this case, that base cost is $50,000, while the cost per unit is $40. So:

$$y = 40x + 50,000$$

17. D: Area = length x width. The answer must be in square inches, so all values must be converted to inches. $\frac{1}{2}$ ft is equal to 6 inches. Therefore, the area of the rectangle is equal to:

$$6 \times \frac{11}{2} = \frac{66}{2} = 33 \text{ square inches}$$

18. D: There are no millions, so the millions period consists of all zeros. 182 is in the billions period, 36 is in the thousands period, 421 is in the hundreds period, and 356 is the decimal.

19. A: Every 8 ml of medicine requires 5 mL. The 45 mL first needs to be split into portions of 8 mL. This results in $\frac{45}{8}$ portions. Each portion requires 5 mL. Therefore:

$$\frac{45}{8} \times 5 = \frac{45*5}{8} = \frac{225}{8} \text{ mL is necessary}$$

20. A: o. The core of the pattern consists of 4 items: ▲oo□. Therefore, the core repeats in multiples of 4, with the pattern starting over on the next step. The closest multiple of 4 to 42 is 40. Step 40 is the end of the core (□), so step 41 will start the core over (▲) and step 42 is o.

21. A: If each man gains 10 pounds, every original data point will increase by 10 pounds. Therefore, the man with the original median will still have the median value, but that value will increase by 10. The smallest value and largest value will also increase by 10 and, therefore, the difference between the two won't change. The range does not change in value and, thus, remains the same.

22. A: Lining up the given scores provides the following list: 60, 75, 80, 85, and one unknown. Because the median needs to be 80, it means 80 must be the middle data point out of these five. Therefore, the unknown data point must be the fourth or fifth data point, meaning it must be greater than or equal to 80. The only answer that fails to meet this condition is 60.

23. A: Let the unknown score be x. The average will be:

$$\frac{5 \cdot 50 + 4 \cdot 70 + x}{10} = \frac{530 + x}{10} = 55$$

Multiply both sides by 10 to get $530 + x = 550$, or $x = 20$.

24. C: Line graph. The scenario involves data consisting of two variables, month and stock value. Box plots display data consisting of values for one variable. Therefore, a box plot is not an appropriate choice. Both line plots and circle graphs are used to display frequencies within categorical data. Neither can be used for the given scenario. Line graphs display two numerical variables on a coordinate grid and show trends among the variables.

25. C: A die has an equal chance for each outcome. Since it has six sides, each outcome has a probability of $\frac{1}{6}$. The chance of a 1 or a 2 is therefore:

$$\frac{1}{6} + \frac{1}{6} = \frac{1}{3}$$

26. B: According to the order of operations, multiplication and division must be completed first from left to right. Then, addition and subtraction are completed from left to right. Therefore:

$$9 \times 9 \div 9 + 9 - 9 \div 9$$

$$81 \div 9 + 9 - 9 \div 9$$

$$9 + 9 - 9 \div 9$$

$$9 + 9 - 1$$

$$18 - 1 = 17.$$

27. B: $12 \times 750 = 9,000$. Therefore, there are 9,000 milliliters of water, which must be converted to liters. 1,000 milliliters equals 1 liter; therefore, 9 liters of water are purchased.

28. D: The two lines are neither parallel nor perpendicular. Parallel lines will never intersect or meet. Therefore, the lines are not parallel. Perpendicular lines intersect to form a right angle (90°). Although

the lines intersect, they do not form a right angle, which is usually indicated with a box at the intersection point. Therefore, the lines are not perpendicular.

29. A: A common denominator must be found. The least common denominator is 15 because it has both 5 and 3 as factors. The fractions must be rewritten using 15 as the denominator.

30. B: An equilateral triangle has three sides of equal length, so if the total perimeter is 18 feet, each side must be 6 feet long. A square with sides of 6 feet will have an area of $6^2 = 36 \; square \; feet$.

31. C: 216cm. Because area is a two-dimensional measurement, the dimensions are multiplied by a scale that is squared to determine the scale of the corresponding areas. The dimensions of the rectangle are multiplied by a scale of 3. Therefore, the area is multiplied by a scale of 3^2 (which is equal to 9):

$$24cm \times 9 = 216cm$$

32. D: Let a be the number of apples and b the number of bananas. Then, the total cost is $2a + 3b = 22$, while it also known that:

$$a + b = 10$$

Using the knowledge of systems of equations, cancel the b variables by multiplying the second equation by -3. This makes the equation:

$$-3a - 3b = -30$$

Adding this to the first equation, the b values cancel to get $-a = -8$, which simplifies to $a = 8$.

Estimation Questions

33. C: To estimate the sum, the numbers after the decimals can be ignored since they are both less than 0.5. This means that the we can round down and then just add the integer:

$$
\begin{array}{r}
3 \\
2 \\
+4 \\
\hline
9
\end{array}
$$

34. D: The fraction is converted so that the denominator is 100 by multiplying the numerator and denominator by 4, to get:

$$\frac{3}{25} = \frac{12}{100}$$

Dividing a number by 100 just moves the decimal point two places to the left, with a result of 0.12.

35. B: 30% is 3/10. The number itself must be 10/3 of 6, or:

$$\frac{10}{3} \times 6 = 10 \times 2 = 20$$

Alternatively, 30% is about $\frac{1}{3}$, so 6 x 3 = 18. The solution would need to be close to 18, making 20 the best choice.

36. B: To make mental math possible, the numbers under the square root sign can be rounded. 8.2 can be rounded down to 8 and 5.75 can be rounded up to 6. 8 squared is 64, and 6 squared is 36. These should be added together to get $64 + 36 = 100$. Then, the last step is to find the square root of 100 which is 10.

37. A: The easiest way to estimate the answer to this problem is changing $800 \div 9$ to $800 \div 10$, which equals 80. This makes Choice *A*, 90, by far the closest option.

38. A: This problem can be estimated by rounding 5.88 up to 6 and then multiplying. $6 \times 3 = 18$.

39. C: If Danny runs 3 miles in 35 minutes, he runs one mile in approximately 12 minutes, since:

$$3 \times 12 = 36$$

Then, to determine how long it will take him to run 5 miles at the same pace, 5 is multiplied by his rate for one mile (12 minutes).

$$12 \times 5 = 60 \; minutes$$

40. D: $27\frac{3}{4}$ inches can be rounded to 28 inches to make the calculation easier. Because Denver got about 3 times as much as New York City (which we are rounding to 28 inches), the 28 inches of snow must be multiplied by 3, which is 84 inches. To make the mental math even easier, 28 inches can be rounded to 30, which, when multiplied by 3 is 90 inches and then the 6 inches can be subtracted (2 for each time 28 was increased to 30), which again yields 84 inches.

41. C: The sum total percentage of a pie chart must equal 100%. Since the CD sales take up less than half of the chart and more than a quarter (25%), it can be determined to be 40% overall. This can also be measured with a protractor. The angle of a circle is 360°. Since 25% of 360 would be 90° and 50% would be 180°, the angle percentage of CD sales falls in between; therefore, it would be answer *C*.

42. B: Since $810 is the price *after* a 20% discount, $810 represents 80% of the original price. This price can be rounded down to $800 for easy mental math since $800 is 80% of $1000. That means that Frank saved about $200 from the original price since $1000 - $800 = $200.

43. D: The total faculty is 14 + 20 = 34. So, the ratio is 34:200. 34 is approximately 1/3 of 100 (since 1/3 = 33.3), so 34 would go into 200 about six times (2×3), yielding the result of 1:6. Choice *C* is incorrect because it reverses the order of the requested ratio.

44. D: 90% of 20 can be quickly calculated by solving 90% of 10 and then multiplying that answer by 2. 90% of 10 is 9 and:

$$9 \times 2 = 18$$

45. D: This problem can be solved by rounding the number of pages read in four nights down from 21 to 20, making the division by 4 simple. 20 pages read over 4 nights is equal to:

$$20 \div 4 = 5 \text{ pages per night}$$

102 pages can be rounded down for easy mental math to 100, so if she reads 100 pages at her rate of 5 pages per night:

$$100 \div 5 = 20 \text{ nights}$$

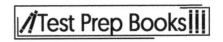

46. B: First, the weight loss in pounds is calculated: 200 − 175 = 25 pounds. Using the conversion rate, the projected weight loss of 25 lb. can be multiplied by 0.50 (rounding up from the given 0.45 $\frac{kg}{lb}$) to get the amount in kilograms (12.5 kg).

47. D: Simply by eyeballing the monthly earnings ($2434.50) and expenses ($1437), it can be seen that the income exceeds the outcome by almost exactly $1000. Therefore, after 3 months, Johnny will have saved three times this amount, or $3000.

48. B: Dividing by 98 can be approximated by dividing by 100, which would mean shifting the decimal point of the numerator to the left by 2. The result is 4.2 which rounds to 4.

49. B: To solve this correctly, keep in mind the order of operations with the mnemonic PEMDAS (Please Excuse My Dear Aunt Sally). This stands for Parentheses, Exponents, Multiplication, Division, Addition, Subtraction. Taking it step by step, solve the parentheses first:

$$4.2 \times 7 + (4)^2 \div 2$$

Then, apply the exponent:

$$4.2 \times 7 + 16 \div 2$$

Multiplication and division are both performed next, but to do so easily, 4.2 can be rounded to 4:

$$28 + 8 = 36$$

Addition and subtraction are done last. The solution is 36.

50. D: Three girls for every two boys can be expressed as a ratio: 3:2. This can be visualized as splitting the school into 5 groups: 3 girl groups and 2 boy groups. The number of students which are in each group can be found by dividing the total number of students by 5:

600 divided by 5 equals 1 part, or 120 students per group

To find the total number of girls, multiply the number of students per group (120) by how the number of girl groups in the school (3). This equals 360, answer *D*.

Ability

Figure Matrices

1. C: The rule for the rows (from left to right) is that ¼ of every shape is taken away, starting from the bottom right and moving clockwise.

2. B: Let's look at the rule for both the row and column in this matrix. The row (left to right) has the exact same bottom shape in each box. The only difference is that the third bottom shape is not shaded in like the first two. Therefore, we can see that the bottom shape is a non-shaded hexagon, leaving us with *B* or *D*. Now let's look at the rule for the column. The top shape is the exact same figure all the way through, even its shading. So, the top shape of our answer is going to be a shaded trapezoid. Therefore, the correct answer is Choice *B*.

3. C: First let's look at the rows going from left to right. We can see that the same figure is represented horizontally through all three rows. For example, if you place your finger on the circle in the first box and drag your finger right, the circle is represented in the same horizontal plane in all three boxes. Therefore, we know the middle figure in our answer will be a square. We can also see, going from row to row, that the bottom figure is the same in box 1 and 3, and the top figure is the same in boxes 2 and 3. Therefore, we can add the bottom shape of the first box and the top shape of the second box to our box in question. In doing this, we see that Choice *C* is the correct answer.

4. B: In this matrix, we can see the pattern best when going from left to right in the rows, so let's look at what the rules are. Every box has a circle inside it. Each box has a shaded line that moves to the right clockwise, and each box has a non-shaded line that stays exactly the same from left to right. In the box in question, we know that the shaded line will be pointing upwards toward the right (since the previous box has it pointing straight up), and we know that the non-shaded line will stay pointing towards the bottom right, so our missing piece is Choice *B*.

5. D: In this one, look at the rules for the columns (going up and down). We know that the middle shape is the exact same in each column, so we are going to have a plain circle, which eliminates Choices *B* and *C*. Next, we know that the middle figure alternates in shading. Therefore, the middle shape is going to be a shaded circle, Choice *D*.

6. C: In this matrix we can see that, in both the column and the row, we are going to have one of each shape. Let's look at the last column. We have a hexagon, a circle, and the question mark. Each of the other columns has (in no particular order): a circle, a hexagon, and a square. So, we know that we are looking for a square, which eliminates Choices *B* and *D*. Now we know we are looking for a square, but will it have stripes going to the right or the left? The other figures with stripes are going toward the left, so we should pick the same one in that pattern. Choice *C* has a square with stripes going toward the left, so this must be our correct answer.

7. A: We can best figure this one out by looking at the rows going across the matrix. In our final result we know that there should be three shapes inside each other. In the first two rows we know that the shape in the first box will be represented as the biggest shape in the third box. In the third row, this will be a circle, so Choices *C* and *D* are eliminated. We also can see that the middle shape in the figure will be the second box in the row, so Choice *B* is eliminated, leaving us Choice *A*.

8. B: In this matrix we can see that the shape is the same in all three rows, thus we eliminate Choice *A*. We also can see in column three that the shape will have three dots associated with it. Since Choice *C* only has two dots, we will eliminate that answer, leaving Choices *B* and *D*. To choose between these two shapes, we have to look at the rows going from left to right. We see that the shape will turn counterclockwise in each box, so the third box will feature the shape upside down. Therefore, the upside-down triangle in Choice *B* is the best fit here.

9. D: Going from left to right in the rows, we can determine that the checkered shape is the same in all three boxes. Therefore, we are looking for a checkered circle, which eliminates Choices *B* and *C*. We can also see that the shape within the checkered circle must be a hexagon since it's the same shape in all the other rows. Therefore, Choice *D* is the best answer here.

10. C: From left to right, we know that all three boxes have the same outside shape. In the third row, this shape is a hexagon with a pointy top, with eliminates Choices *B* and *D*. We also can see that the middle shape and the very center shape are the same too. In the third row, we know we are going to have a circle inside a trapezoid inside a hexagon, so Choice *C* is the correct answer.

11. C: Going from left to right, we see that the basis of each row is the same shape with a cross or a plus in the middle. In the first box there is always one dot, in the second box there is always two dots, so there must be three dots in our unknown box. Notice that the third box in rows one and two adds a third in the middle, but the two on the outer portion of the box do not change. We should match this same pattern with our answer; C is the best answer because the two dots remain on the top and on the right side in this answer choice.

12. B: For this matrix we can eliminate Choice C because we know the outer shape in the boxes in row three are going to have seven points, and Choice C has nine points. We also know that the inner shape is going to be a hexagon, as determined by the other rows and their consistent middle shapes, so this eliminates Choice D. Finally, we can eliminate Choice A because it does not have a third shape around the inner shape (which should be a square). Box three in all the rows has three shapes, and Choice A only has two shapes.

13. D: Notice that each row in this matrix has a different pattern in the inner circle. There is at least one with an "X," one with a "+" and one with eight lines in the very middle. In row three, we already have one with a "+" and one with an "X," so we need the one with the eight lines in the middle, which is Choice D.

14. C: Each row in this matrix utilizes the same shapes in all three boxes with the same shading, so we can eliminate Choice A because it alternates in shading. Choice D can also be eliminated, because in the third column we see that the outer shapes have no triangles sticking out of them. That leaves Choices B and C. If we look at the outer pentagon shape in the third row, we see that the point is sticking straight up to the top, so Choice C is our best answer here. Choice B is tilted to the right and does not reflect the other two shapes in the row.

15. A: Let's look at the first column. We see that the shaded trapezoid is in all three boxes in column 1, and that it moves around to either the top, middle, or bottom of the box. It's the same with the striped square and the lightly-shaded hexagon. In column 2, we see that there is a shaded square that moves around from spot to spot, a striped hexagon, and a lightly-shaded trapezoid. Now we know in column 3 there will be the same shaded-in shapes. This means our final box will have a lightly-shaded square, which eliminates Choices B and C. Now let's look at the placement of the shapes. The lightly-shaded square in column 3 is already represented at the top and middle, so we need a lightly-shaded square at the bottom of the box. Choice A reflects this pattern perfectly.

16. D: In each row of this matrix (left to right), we see the original shape in the same position, but there is always one extra line with a dot added to the bottom. So for the third row, we want to find the shape in box 1 apparent in the shape in the answer, with two lines added to it. We see in Choice D that this shape is facing the same direction as the boxes in row 3. The only difference is that there is a line added to the bottom of the shape. Therefore, our best answer is Choice D.

17. B: The easiest way to determine the missing box is by the columns (up and down). In the first column, we see that in the first and third box, the arrows are flipped. It's the same with the second column. If we flip the pair of arrows in the first box, we get what we have in the third box. So, we must take the first box in column three and flip it. We would have the top arrow pointing to the bottom left and the bottom arrow pointing to the top right, as we see in Choice B.

18. A: Let's look at the rows across first. We find the middle (second) box by adding the first and third box together. In the first row, we add the square and the "X" together with the shape with the square and circle in the middle, and we get the middle image. The same happens in the second row, when we add the first and third images together. Now let's add the first and third boxes from the third row. We know from the middle image that we have a star shape inside a square with a circle in the middle. However, we are missing the "X" and the square within the star shape. Therefore, we need to look for an "X" with a square in the middle, which is Choice *A*.

19. B: In this matrix, let's look at the rows across. In the first row, to get the image in the third box, we add box one and box two together. In the second row, we would also add box one and two together to get box three. Let's add box one and two together in the third row. We should have a large shaded circle, four dots, a checkered hexagon, a white square, and then a smaller white circle with four points coming out of it. This describes Choice *B* perfectly, so mark it as the correct answer.

20. D: Let's look at the columns first and determine what the inner shape is going to be. In column one, we see that the inner shape is a triangle. In column two, we see that the inner shape is a square. In column three, we see that the inner shape is a star. Therefore, we can eliminate Choice *C*. Now, let's determine the shading of the outer shape. We see that each row has a light gray shade, a dark gray shade, and a white outer shape. In the third row, we are missing the white outer shape. Therefore, that narrows our answers down to Choices *B* and *D*. The only difference between these two is the way the inner shape is tilted. In the first and second column, the first and third inner squares have the same tilt, so the first and third inner shapes of the third column should also have the same tilt.

Paper Folding

1. D: The paper is folded in half, creating two layers. Three holes are punched through the two layers. We see the holes mirrored in Choice *D*. The top hole is near the center, so we know that the top hole on the right side will also be near the center.

2. C: The paper is folded in half, creating two layers. Choice *C* is the best answer because the left side is mirrored perfectly over on the right side. Notice how the triangle facing up is mirrored by another triangle facing up, and the triangle pointing right is mirrored by the triangle pointing left. If we draw a line down the middle of Choice *C*, we see how the two sides reflect each other.

3. B: The paper is folded in half, creating two layers. Then it is folded again, creating four layers. Let's flip the fourth of the paper *down* first. We know that the triangle at the bottom will be represented right next to itself, so we will see triangle, triangle (left), circle, like in Choice *B*. Choice *D* is incorrect because the triangle at the bottom of the fourth is pointing in a different direction than the original image.

4. A: The paper is folded into fourths. Let's fold it back over to the left first. We know that the circle is going to be closest to itself, so we'll have two circles right at the top middle, eliminating Choice *B*. Next, we know that when we fold it over we will have two circles in the very middle, because the circle closer to the edge will produce another circle close to the edge, eliminating Choice *C*. Choices *A* and *D* look the same when folded over into the half, so let's unfold it all the way. We know that when the paper is shown as a half, there is a circle close to the middle, so when we unfold the paper all the way, there will be *two circles* in the very middle mirroring each other. Choice *D* does not have this mirroring effect, so Choice *A* must be the correct answer.

5. D: The paper is folded into fourths, and we have to unfold it up first. We see that the circle at the top will be recreated as two circles in the middle. This eliminates Choices *A* and *B*. The only difference between Choices *C* and *D* is that the top star either has its top point facing up (Choice *D*) or facing down (Choice *C*). Look how the point of the star is facing straight downward in the original. If we flipped the fourth up, the star would be facing straight upward. Therefore, Choice *D* is the best answer.

6. A: Let's eliminate Choices *B* and *D*, because if we flipped the diagonal paper over, the bottom shape would be mirrored, thus creating the large shape in Choices *A* or *C*. If we flipped the paper over, the two triangles at the left top would be flipped, but they would still be pointing toward the right at the bottom of the page, so Choice *A* is the best answer.

7. D: This one looks difficult at first, but let's take a look at the original marking. There is a diamond in the middle, two circles above the diamond, and a circle on the right and left side of the diamond. All we have to do is make sure this pattern is recreated in one of the answers. The only answer this is recreated in is Choice *D*, so this is the best answer.

8. A: Again, this one looks difficult but it's simple if we just look for the original image. Choice *A* fits perfectly but let's rule out the others. Choice *B* is incorrect because one of the triangles is facing right instead of left. Choice *C* is incorrect because there is no straight line to the very left; rather, the line is represented in the middle. Choice *D* is incorrect for the same reason as Choice *C*.

9. B: For this, we simply fold down the edges of the paper. If there is a hole punched at the very edge, there will be three more holes reproduced when the paper unfolds. Therefore, Choices *C* and *D* can be eliminated. Choice *A* is incorrect; if we look at the original, we see that the squares at either end of the paper are in line with the middle square. This does not happen in Choice *A*. Therefore, Choice *B* is the best answer.

10. C: The paper here is folded up at the corners, the holes are punched, and then the corners are folded back down. Choice *A* cannot be the correct answer because if you look at the bottom triangles, the one to the right is facing the opposite direction of the original. Choice *B* is incorrect because of the same reason; two triangles that are supposed to be facing right are actually facing left. Choice *C* looks like the best answer. Choice *D* is incorrect because if we flipped the bottom corner down, the triangle would mirror itself, and instead toward the left we have two triangles facing up, which is incorrect.

11. D: We can immediately eliminate Choice *B*. If we fold the left top and right bottom folds out, we would have a single dot in the corner followed by two dots after. Now let's look at Choice *A*. Choice *A* at the bottom left is missing the dot in the very corner that would be produced if the triangle were folded back out. Choice *C* has the dot in the bottom left; however, it has an extra dot right behind the corner dot that should not be there. Therefore, Choice *D* is our best answer.

12. C: All four corners are folded up on this paper, hole-punched, then folded back down. Notice that if you curl the edges up in a triangle, there would be one small triangle in the very corner. Thus, we can eliminate Choices *B* and *D*. The difference between *A* and *C* are the positions of the corner triangles. Let's look at the top right corner. *A* is facing up and *C* is facing down. If we mirror the corner triangle in the original, the mirror would be facing down in the top right, not up. Therefore, Choice *C* is our best answer.

13. B: This one is tricky. Let's say we fold it down once and then hole punch it. Then we would have three punches up top. But say we have that top part folded then fold it over to the right. This means the top corner would also be included in the hole punch. Therefore, that's how we get the top left punch in the corner, along with the mirrors in the original.

14. D: Since we've folded the top left corner down and over, and we punched the edge, we know we are going to have a triangle in the top left corner, thus eliminating Choices *B* and *C*. Let's look at the original. Now say we punch the upside-down triangle at the top left, and then unfold the two sides over and up.

We would get a *right side up* triangle as its mirror. This is depicted in Choice *D*, but not in Choice *A*, so *D* is the correct answer.

15. A: Let's look at the top left again for our correct answer. If we fold the top down, the left side over, and then punch a hole in the corner, we know we are going to have a punch at the very left top. Thus we can eliminate Choices *B* and *C*. We know that the punch in the top left corner in the *original* paper is going to be mirrored when we unfold it to the left, so it'll be facing *itself*, toward the right. Choice *A* effectively demonstrates this mirroring.

16. B: Right away we can eliminate Choice *A* because the two holes at the top of the original paper are not mirrored in the center. Likewise, we can eliminate Choice *D* because the original two holes at the bottom are not mirrored to the right. Now take your finger and put it diagonal over Choice *B*. If you look at it cut in half diagonally, we can see that the two sides are perfectly mirrored. Do the same for Choice *C*. The two are not mirrored perfectly diagonally. In fact, there's a hole in the top left corner that should not be there at all.

17. C: Look at the original paper at the very top. We see two triangles pointing toward the left. Once we flip that side back up, the two triangles will still be pointing to the left. With this information, let's eliminate Choices *A* and *B*, the latter of which has no flipped triangles. When we look at the triangles diagonally they should be inverted, which is Choice *C*. In Choice *D*, when flipped, the triangles are facing the opposite way.

18. A: Let's look at the original paper. The first thing we should do is flip the left side of it back up. We will expect to see a mirrored triangle. Right now the triangle on the top left is pointing upside down, but when we flip it, it should be facing right side up. The only answer choice that has this possibility is Choice *A*, so this is the best answer.

19. D: When looking at the original paper, let's flip the first side up. We should have two whole squares and a square underneath it, which eliminates Choices *B* and *C*. When we have a half circle paper, we should have four whole squares and two half squares, which makes 5. To find out the correct answer, let's double the amount of squares by unfolding the paper all the way. We should have ten squares altogether (5 + 5). Thus, Choice *D* is the correct answer, because it is the only one with ten squares.

20. C: Unfolding the original paper, we know that we will have a triangle pointing at itself in all four centers. So, we can eliminate Choices *A* and *D*. We also know that we will have the triangle that faces up face down on itself, which is only true in Choice *C* out of the Choices *C* and *D*. Therefore, the best answer is Choice *C*.

Figure Classifications

1. C: All figures are triangles.

2. D: All figures are a circle within another circle.

3. B: All figures are shaped like diamonds with stripes inside them.

4. E: All figures have one tail on them.

5. D: Each figure is a square with two squares inside of it, with three squares in total.

6. C: Each figure is a double-lined square with four triangles inside of it.

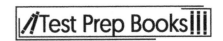

7. B: Each figure has four sides with a cross inside it.

8. D: All figures have four sides with a cross and a box inside them.

9. C: Each figure has five sides.

10. E: All figures are circles.

11. B: Each figure is an arrow and each is pointing either up, down, left, or right.

12. A: There are three figures inside each other. The first is a hexagon, then a circle, then another hexagon.

13. D: Each figure adds a line. The fourth figure must have four lines, like Choice *D*.

14. B: Each has the same shaded pattern. They all have a white sliver, a lightly-shaded sliver, and a dark sliver.

15. E: Each figure has a hexagon as the outer shape, then a circle with 8 slivers, then a circle in the middle. Choice *E* is the only other shape that matches this description.

16. C: All shapes have nine points, then a triangle that is outlined on the inside.

17. D: Each is a checkered triangle with a circle in the middle. The first three triangles have a circle up top, on bottom, and to the left, so we are looking for a triangle with a circle to the right. Choice *D* fulfills these criteria.

18. A: Each shape is a circle with three lines toward the top. Choice *A* is the best answer. Choice *B* has three lines, but the lines are toward the bottom instead of toward the top.

19. E: All figures are triangles with a circle inside facing a certain way. We are missing a circle facing down, so Choice *E* is the best answer.

20. C: We see that the shape curves inward with every new shape, making its way to a star shape.

21. E: Each figure has four sides. The lines within the figures are curvy, and each has three dots on the outside and two on the inside of the lines, a total of ten dots. Choice *E* fits all these criteria.

22. C: With each of these figures, they all have arrows pointing out of all four sides. The only other shape with *all* arrows is Choice *C*.

23. B: These are all star-shaped and have long points on them. The only other shape like this is Choice *B*.

Dear TACHS Test Taker,

We would like to start by thanking you for purchasing this study guide for your TACHS exam. We hope that we exceeded your expectations.

Our goal in creating this study guide was to cover all of the topics that you will see on the test. We also strove to make our practice questions as similar as possible to what you will encounter on test day. With that being said, if you found something that you feel was not up to your standards, please send us an email and let us know.

We would also like to let you know about other books in our catalog that may interest you.

ACT

This can be found on Amazon: amazon.com/dp/1628458844

SAT

amazon.com/dp/1628458984

SAT Math 1

amazon.com/dp/1628458631

ACCUPLACER

amazon.com/dp/162845945X

TSI

amazon.com/dp/162845721X

AP Biology

amazon.com/dp/1628456221

We have study guides in a wide variety of fields. If the one you are looking for isn't listed above, then try searching for it on Amazon or send us an email.

Thanks Again and Happy Testing!
Product Development Team
info@studyguideteam.com

FREE Test Taking Tips DVD Offer

To help us better serve you, we have developed a Test Taking Tips DVD that we would like to give you for FREE. **This DVD covers world-class test taking tips that you can use to be even more successful when you are taking your test.**

All that we ask is that you email us your feedback about your study guide. Please let us know what you thought about it – whether that is good, bad or indifferent.

To get your **FREE Test Taking Tips DVD**, email freedvd@studyguideteam.com with "FREE DVD" in the subject line and the following information in the body of the email:

      a. The title of your study guide.

      b. Your product rating on a scale of 1-5, with 5 being the highest rating.

      c. Your feedback about the study guide. What did you think of it?

      d. Your full name and shipping address to send your free DVD.

If you have any questions or concerns, please don't hesitate to contact us at freedvd@studyguideteam.com.

Thanks again!

Made in the USA
Monee, IL
01 October 2021